AF339064

VEGETARIAN GEOLOGY

IS IT A TRUE CONCEPT?

By ALLEN LINDERMAN

FIRST EDITION
1980
SECOND EDITION
1981
THIRD EDITION
1982

Library of Congress No. 82-061274
ISBN No. 0-87970-154-4
Copyright© 1980, 1981, 1982 by
ALLEN LINDERMAN

First Printing 1980
Second Printing 1981
Third Printing 1982

Printed by:

North Plains Press
Aberdeen, S.D. 57401
Printed in U.S.A.

Allen Linderman
R.R. No. 1, Box 131
Carrington, N.D. 58421

This is a U 15 drilling rig on lease working in Williams County, North Dakota. It has a 127-foot mast and will drill over 9,000 feet. This well came in flowing as did the offset well. With our geologist and drilling superintendent, it was not a problem to drill our own wells, five were drilled in this area. My job was to supervise drilling operations. The first well drilled here was a wildcat, which opened up a new oil field 1960-61.

AN OBSERVATION

It has been said in modern scientific circles, *"One of the most amazing things is that the universe should be so close to the dividing line between collapsing and expanding."* It is thought a closed universe will pulsate back and forth into different universes. Why must it do either? Is it so unthinkable that velocity and attraction, could not eventually find permanency?

A COMMENTARY

"There is an old saying that one does not convert the opponent of a truth. One waits for them to die, and a new generation to grow up, acquainted with the truth from the beginning." (Dr. Carlton Fredericks)

Sad as this comment may be, it has had all too loud a ring throughout historical times. This will not apply to my intelligent reader and all other intelligent people past and present.

Mr. Allen Linderman   January 22, 1981
R.R. No. 1, Box 131
Carrington, N.D. 58421

Dear Allen:

This is to admit that I read your book — found it very deep — very thoughtful — and certainly different.

I am reading it again trying to grasp an understanding of the magnitude of some of the questions you raise and some of the potential answers.

You are to be congratulated on, not only doing some excellent thinking, but being able to express those thoughts.

Sincerely yours,

Harold

Harold Schafer

Chairman — Board of Directors, Gold Seal Company, Bismarck, North Dakota

PREFACE

The critic may say: This book covers too many points and is not well arranged. Some points may be placed which would fit better somewhere else. I will beat the critic to the punch, by admitting this may be true. Also, the author quotes from many other sources for his material. I do this, as many authors do when writing on such technical material. I like to show who wrote such material, and give opposing viewpoints as well. I also do this and do not apologize.

On the other hand, I am not a professional writer, even with my great knowledge of my subject, I may not always enter material in the best or proper place. For this I do apologize, and hope this will not in anyway diminish from the sound, factual information I am so anxious to share with you. I know, after years of research, all of the information in my book is important and contributes to my main theme. Please do not allow my lack of professional writing ability to detract from the information contained in my book. It is you, my reader, I want to please, and share this very important information on geology, and related subjects.

It has not been my desire to regionalize this book, or to direct my work to only scientists and professional groups. This book will appeal to persons of every walk of life, the young, the student, and all ages and interest in life. You will find it both educational and informative. From the introduction to the last paragraph it will hold your interest. There will be hypothesis you may never have heard of. My wish is to give you an entire new outlook on creative processes and geology. It is not difficult to understand and with the use of a dictionary at times, this book will enrich your life. You need not be a paleontologist to clearly understand the progression of thought presented here, and it will unravel old, hard to understand, teachings. The footprints of archaeology and ancient history will be pieced together like a jigsaw.

Because of having been born and reared in North Dakota much information may be centered around the area and state where my family have lived for about four generations (though I have traveled extensively and lived other places including California). We must keep in mind the geological stratigraphic column is earth wide and not regional. Had I been born in one of the other great oil producing states and there are many, such as Texas, Louisiana, Oklahoma, Kansas, Wyoming, Montana, or California, I may have centered around these areas. This, however, would not change far reaching geological development. This is not a regional book, it is a worldwide book. It will be read and used as a reference book five or ten years from now.

Because much of the thought and ideas presented here are so different and contrary to that taught in the Western World for the last 100 years — technology and evolutionary science; in that we have simply inherited Darwinism and Mendelian (Gregor Johann Mendel-1822-84, a monk — Mendel's law of genetics, Brno, Czechoslavakia) theory on biological science, I feel I should give you a few main points and terms at the outset so you will be alert to catch them readily and enrich your reading time. Some new points and terms are:

Oil, coal, and carbon are not made of organic plant and animal life. The process is far too puny.

Carbon content of the earth is so great it is not calculable. Much may never be recoverable. This fact will not solve our energy problems.

Carbons were distilled in the original hot igneous earth long before plant or animal ever lived. There is a carbon and mineral ring all the way around the earth, even under the oceans. As on land, not contiguous.

Organic substance will not become carbon in the presence of free oxygen. Hydrogen and carbon will not become oil in the presence of free oxygen.

Oil, and/or gas, are now found nearly six miles deep, this is below fossilization. Carbons are known to be on other planets without life on them.

The word "annular" will be used many times in this book. The "earth's annular system" means, the earth once had its own ring system just as Saturn and some other planets still do. Annular geology is profoundly different from old-school geology; yet, it answers all of the unanswered questions, not answered by traditional geology.

The material for ring systems on planets arose from the planets themselves. One main proof for this is: All ring systems on planets rotate at the equatorial plane. Other material may or may not rotate at the equatorial plane. See the picture in this book, Early Stage of Ring Formation, page 238.

When the earth was very hot, boiling and igneous, it threw much of its crust material, as we know it, into a ring (annular) system around itself. Later, after cooling, this material fell back to the earth in a segregated order because of gravity and centrifugal law.

Water, carbon, and minerals went up into this ring system. As with Saturn today, much of the water remained in the outer ring, only to fall last. This built the polar ice caps, as it seldom if ever snows on either pole. Therefore, the earth was stratified from the bottom up, not the top down. Water is found as deep as 16,000 feet. It was placed there when the formation containing it was. (Page 239)

The word "augmentation" is used many times. The oceans of the world have been augmented with water over millions of years time from annular rings. Isostasy, which is hydrostatic compression against and around all of the continents of the world, caused by oceans and their augmentation, is a provable concept. This heaved the continents higher.

Continents were continually heaved higher because of ocean augmentation and isostasy. If you were to place a pan of water on a stove burner and turn it on high long enough, every drop would steam away into the atmosphere and the pan would be destroyed. Therefore, in the early stages of earth development with such heat, every drop of water on earth must have been thrown into outer space as rings around the globe, later to descend.

Slow creeping glaciers did not formulate the earth's topography as we know it today. Great down-falls of water and fluvial actions built the irregular earth. Erosion would build only flat lands and plateaus. Polar ice caps are right where they fell from the ring system (annular). Glaciation of the earth was caused by water falling from its ring system, not icebergs sliding down from the north with no energy force to move them.

What is thought to be metamorphism is blending. (Metamorphose, change in mineralogical structure.) Metamorphism is the great geological fiasco.

Elemental segregation and assortment was done before placement in the earth, not after. The great iron and metallic beds of the earth were of annular development. Because of annular placement, metallic beds are, and will be, found in most of the oceans of the world.

A professor of geology from the University of Minnesota wrote after reading my first book: *"I must admit that I was not familiar with many of your hypothesis, but I find them interesting and refreshing to read. It is people like yourself who keep scientists on their toes by reminding them that there is more than one side to scientific discourse. I think that is a good thing."* An independent oil Landman wrote: *"I found your book, 'Vegetarian Geology, Is It A True Concept?', both interesting and informative . . . I would like to add that previous to reading your book; through my travels in the world; I, too, saw in my opinion, what was evidence of a global flood throughout the various geological formations, I have contested the thought of a giant ice age that once covered the Earth millions of years ago . . . The "annular Theory" does not surprise me, I do thank you for bringing new light to an old subject."*

When autographing books at a B. Dalton Bookseller store, an Air Force pilot bought a book and said, *"He had flown the western United States for years and as he looked at and studied the earth, he was convinced its shape was caused by a great flood of water."* At another B. Dalton store, a man who said he had drilled oil wells in the Middle East, South America, Texas, and other places and now in North Dakota, and in his expert opinion, oil and gas as well as coal, are not made of marine life and vegetation.

A man from Indiana wrote: *"I have received and read your book. I have already begun re-reading it, and quite probably will again several times. . . . It was a mind opening experience with a great deal of 'food for thought'. The main problem with the treatise (my first book) is that it leaves me hungry for more. . . . You are my first and only contact with some of these concepts. I truly hope to learn more."* Another man from Illinois wrote: *"Have just finished reading your book on Vegetarian Geology, and found it very interesting and informative. I am a self-made amateur geologist with no formal education in geology and I must say you have changed many of my thoughts to a more realistic point of view. . . . I thoroughly enjoyed reading your works, keep up the constructive progress."* Someone from California wrote: *"I have very much enjoyed your book 'Vegetarian Geology' and profited thereby. . . . The press generally assumes continental drift and plate tectonics, erosion and a lot of other nonsensical theories about the creation. I thank you for having gone to the trouble of writing this book which I bought at Knotts Berry Farm in the Rockshop. . . . Am writing to ask for another copy as I am sending mine to a friend."*

When I was a boy I began to think along these lines. I have never been a conformist. Proving someone wrong and myself right is not the objective of this material or my book on Vegetarian Geology. After a lifetime of study, research and thought, I have a sincere desire to share what I have learned with others. This is the first book of its kind in over fifty years, perhaps eighty.

INTRODUCTION

This book, or thesis, will not be what you may think it is from the title. It is different in many ways and you will need to read it to learn the many different subjects. It is my wish and desire to bring interesting, educational and valuable information to the reader. It will challenge many long standing theories and dogmas. This is not done to be radical, but rather to show, by reason, why we cannot accept every teaching, no matter how old.

Most authors start out with a long list of schools and universities, and research fields, they have. I do not have these to fall back on, though my mind is not regimented or taught to conform because of them. Freedom to think logically is more important. In place of higher education, I have a world of experience in life. My uncle once said when new ideas come up, most people would say it will never work, and while they are wasting their time arguing about its failure, some of us were out doing it and it did work. Bill Lear was such a man, in electronics and aviation.

There are endless types and kinds of people and we all are needed. To me, though, we may divide them into two major groups. The largest group being the spectator, with the balance being participants. It is fun to be a spectator at a ballgame, but the players have the most fun and make the most money.

My main subject is vegetarian geology. It is not a true concept as I will show. The ice age did not happen as is claimed. Man has not been on earth millions of years. Experts in fossil research are mistaken on the life span of them. Plate tectonic and global plate tectonics, as explained by science, are not correct. The theory of continental drift is still unprovable. There had to be ocean augmentation, which answers a host of problems, as well as made the Grand Canyon, not erosion. The earth's surface, as we now know it, was caused by fluvial actions, not ice. There had to be a global deluge to accomplish the great task of shaping the earth. Vast amounts of energy were required, lacking in slow erosion factors. Earthquakes today still draw on this original energy factor of sudden deluge. Energy for an earthquake originated in the igneous earth.

In any field of scientific research and technology we must consider the rule of factor, is it constant or changing. Think of the law of electrical wavelength or frequency, an absolute constant factor from the point of origin. This never changes; therefore, it is possible to transmit television pictures from other planets to earth. Who caused this unfailing law? What would happen if it failed? Apply this to sound and light.

Many subjects have been brought into the material in order to round out and give substance to the subject. This is necessary to prove and establish a basis for such controversial thought. To be right by accident or just guess right has never appealed to me. Opinion is of no value unless there are some real basic facts to support it. To believe because everyone else does seems to be a waste of time. Accountability seems to be old-fashioned nowadays. Why be encumbered with facts, truth, or proof, which will stand in the way of the objective? Such attitude seems to be the order of our day. Nevertheless, fact will rule our lives regardless of what men may think or do.

So, I do go through a lot of different subjects to build on a main theme, for which I hope you do not mind. As the human body functions as a total unit, and no matter what happens to one part, all of the body is affected. So it is with most subjects, we must consider all of the factors. Centralization is a modern theory and it is not always beneficial. There is too much single dependency of thought. Too often a single line of thought is pursued, when composite factors all relate to the subject. So it is with my thesis.

A GEOLOGICAL PREVIEW

In the rock cycle theory, "new rocks from old", if elemental metamorphism is a continuous factor, there would be an elemental blending to a single substance ultimately, rather than a great separation of rock identity, as we now have. Scientists often do come up with a theory which has some merit, and then destroy the theory by giving it too long a time element. So it is with sedimentation and erosion by wind and rainfall, as we know it to be today. This process would leave the earth's surface more like a peach, than the way it is. Erosion as we know it just is not that old, and is over dramatized. If we assign it millions of years, its cycle would have run its course millions of years ago.

Some geological features even the novice could tell are not caused by erosion are: Owachomo Bridge, in Utah's Natural Bridge National Monument. Rainbow Bridge, southern Utah. All caves and caverns. A huge balanced rock, in the Garden of the Gods, Colorado Springs.

The Lava Pinnacles of Chiricahua National Monument, Arizona. Arizona's Ship Rock, 1,640 feet high. The Devil's Tower, Wyoming. California's Yosemite National Park with majestic breathtaking scenery, where the sheer granite face of El Capitan rises 3,600 feet above the floor, and Yosemite Falls has a double drop of 2,425 feet, where each structure appears to have monolithic majesty. Death Valley with the Devils Golf Course. The Devils Postpile National Monument, California. Spend a few hours on the rim of Bryce Canyon of Utah, and you will agree it was caused by a great fall of water which washed away loose material leaving it as it is today. Read the Reader's Digest book, "Scenic Wonders Of America", then decide for yourself whether erosion or cataclysm built the scenes. Also vertical pillars, wherever they are found, which everyone can see were once part of the plateau. A cataclysm, short in duration, is the only answer to the development of such beautiful scenes. Softer material just washed away, not by millions of years of erosion.

Mountains, shists, folds, overthrusts, upheavals, piercements, tilts, faults, granite slabs, valleys, plateaus, rock stratification, anticlines, synclines, monoclines, saltbeds, coalbeds, magma, islands, and continents, are not built by slow simple erosion. What then has erosion done? True, erosion is a constant factor in our geological times. In recent geological times, perhaps land erosion, river silting, and shoreline erosion, would be the greatest assignment we could give to erosion. It appears reasonable that erosion did not formulate the earth's crust as we see it today. Landslides and rocks falling to lower levels is the work of gravity, as they were not placed there by erosion. We live on a complex and anomalous earth, where everything retains its identity.

No one knows how the universe first originated, the "big bang theory", could be as valid as any. The universe is held together by known and unknown internal forces. With gravity and velocity, and other unknown factors, everything in it will eventually find a permanent home, and will not fall back into a gigantic crash, for the second "bang".

CONTENTS

PART I

VEGETARIAN GEOLOGY AND RELATED SUBJECTS
IS IT A TRUE CONCEPT?
(by Allen Linderman, May, 1982)

CHAPTER 2

CHAPTER 3

CHAPTER 4

CHAPTER 5

CHAPTER 6

CHAPTER 7

CHAPTER 8

PART II
CHAPTER 9 MORE ABOUT GLACIATION

CHAPTER 10

CHAPTER 11

CHAPTER 12

CHAPTER 13

CHAPTER 14

CHAPTER 15

CHAPTER 16

CHAPTER 17

CHAPTER 18
LOOKING BACKWARD AND LOOKING FORWARD

PART I

Traveling east from Yellowstone Park to Cody, Wyoming, there are some very beautiful and outstanding geological formations. First there are towering mountains with rock slabs at the top so thin in places there are see through holes in them. Then jagged and upheaved formations, with great valley gorges which are very irregular. A fantastic scene photographed by many tourists is Chimney Rock in Wapiti Valley west of Cody, shown above. Certainly, erosion is affecting these rugged formations today; however, they were not built by erosion, which would tend to destroy them. These great upheaved structures were built by a great down-rush of water which washed away softer material and left the scene just as we see it today.

CHAPTER 1

VEGETARIAN GEOLOGY AND RELATED SUBJECTS
IS IT A TRUE CONCEPT?
(By Allen Linderman, May, 1982)

An interesting observation on physical law — for every force there is a like force or reaction force equal to the force; therefore, all forces are equal or neutral, or seeking neutrality without choice. When an outside force is applied or a force reduced or increased, all affected positions scatter and seek new positions which, ultimately, will be neutral. Neutrality is the natural position of physical law or force. Everything is either mass or energy — they are interchangeable. Some forms of interference may cause mass to convert to clear energy. Man, in a small way, has the power to do this. God has reserved for Himself the power to cause clear energy to become material mass.

Little is known about true densification, yet we do know a square inch cube of material with zero densification would weigh 100 million tons.

Complex as our universe is, only a few basic physical laws govern. Two laws, time and motion, will take a space flight to the moon and back. There is no time until something moves. One does not exist without the other (they complement), just as material mass is either energy or mass; it must be one or the other.

Just as little is known about gravity, it is only an attraction of mass to mass, in relation to total volume and densification. In reality, how many positions are there in particle physics? Some say no less than two. Yet it has been said that there may be only one and con-science knowledge of the first is the second. Where is oblivion? Only God knows. If we knew where oblivion was, the very knowledge of its location would cancel out its existence. Oblivious makes oblivion.

GRAVITY — ACCELERATION — CONVERGENCE

Will include some information, and questions on the Law of Gravitation. In 1687, Sir Isaac Newton stated a universal law of gravitation. In 1916, Einstein announced his theory of relativity. Both Theories were correct at the time they were released, and have added much to science, though there are some unanswered questions as yet.

We know the point at which the sun pulls on the earth is not the center of the earth. It is the center of the earth and the moon combined. What happens to centrifugal force on the earth's surface from its rotation at the equator, to the poles where it is zero? My encyclopedia says the weight pull of 16 ounces at a temperate zone, which is off the equator, is 15.9 at the equator, then increases to 16.2 ounces at the pole. Centrifugal force makes the weight less at the equator, as this force disappears at the poles. Also polar diameter of the earth is 27 miles less than equatorial diameter. Here we have factors causing the weight increase at the poles. No one yet knows the cause of gravitation. Perhaps this has already been done, if not, may I suggest that a 50 pound substance be weighed at the equator on both a balance scale, and a spring loaded scale (gravity would have the same effect on a balance scale everywhere), then take both the balance scale and the spring loaded scale and the weight to the pole and weigh them there. May or may not prove anything.

This brings us to the actions of the force of gravity and acceleration. Are they the same force? What happens to gravity as it converges from all directions to the center of the earth? Einstein says that objects do not attract each other by exerting a pull. He says curvature causes a gravitational field. Could we then say gravital force of the earth is nonconvergence toward the center of the earth? A question? Who or what is our prime mover? It has been said both gravity and acceleration are vector quantities, consisting of a force and the direction in which the force is acting. Two people standing on opposite sides of the earth would be controlled in two ways by a single force.

The fastest known motion is the speed of light, 186,000 miles a second. Therefore, we know acceleration cannot forever be perpetuated. At optimal, whatever that may be, it is no longer acceleration, yet still requires a prime mover. Hypothetically, if we could fire a rifle bullet straight up, with no wind or outside interference, at the termination of acceleration, wouldn't gravity bring it down to the same exact spot of origin? We could fire the rifle horizontal and the bullet would fall to earth far away in a graduated curve line, in relation to diminished velocity and gravital law. Are the two forces the same? In gravity we have an endless prime mover, though we do not know what it is. With the rifle bullet we know what the prime mover is, duration is short, and nonperpetuating. With acceleration do we not need inertia to accelerate? Does gravity need inertia, or is it just an unknown prime mover? Can gravity be deflected, as acceleration might be? Or may neither be deflected? Someone suggested as a test, a plumb bob in an accelerating elevator will only point down. May I add, if the elevator were standing still the plumb bob would do the same. If the elevator were built 20 degrees off, the plumb bob would point straight down, until optimal acceleration was reached, the plumb bob would make a slight gradual correction, between velocity or acceleration and gravity. As the elevator came to a stop the plumb bob would again point straight down. Would the plumb bob reverse itself with a fast drop to bottom?

In this test did gravity change direction? No, it was only reacting to an outside force, acceleration. Its direction of force was always the same. The question is, did either gravity or acceleration change direction? Let's think about that. See the December 1979, and June 1980, Popular Science for some very interesting thoughts on this subject. As one reader suggested, there is no test that can tell the difference, gravity and acceleration should be compared where the direction can be ignored. The best example I have found was sent in by a reader of Popular Science, June 1980.

. . . *"The apparent dilemma concerning the nonconvergence of accelerational direction can be dissolved if we consider a balloon whose surface is accelerating outward uniformly and in all directions. In this case the acceleration vectors will converge on the center of the sphere. The original problem arose by comparing gravity and acceleration in different geometrical dimensions — planar (one-dimensional) and spherical (three-dimensional). The two conditions are equivalent only when we consider an infinitely small area of a spherical surface."*

My encyclopedia also says: 16 ounces of mass 2,000 miles within the earth weighs 8 ounces, and at the center 4 ounces. I do not question this, although if the law of convergence is applied to gravity, it would seem weight would increase not decrease. What happens to gravity if it does converge from all sides to the center of the earth and collides there? Where does it go if it does not? If gravity gradually fades away to the center of the earth or of any mass, why shouldn't the 4 ounces be zero at the center? Acceleration is gone at optimum speed as men in a space capsule, traveling 26,000 feet per second, float weightless in any direction. Could we say, if there was a small space at the center of the earth, in line with true axis, men or objects could float weightlessly in this space? Would this be true with fade away gravity force, or convergence? What difference would it make? Does anyone really know?

Applying this weight reduction rule towards earth's center, if we had one million pounds of drill pipe on the ground for a 5 mile deep oil well, would it weigh less when it was all in the hole than when on the surface? If our weight gauge on the derrick floor showed one million pounds, we still do not know how much the pipe weighs, as there is water and drilling mud in the hole at the same time. If we had an accurate weight of the pipe on the ground, other factors affect the weight in the hole. Perhaps we have raised more questions than we have answered.

Science has the ability to do many fantastic things, and make great discoveries of already existing and created things, yet not one of them can make a single blade of grass, or ever will.

If there were a time when there was no carbon on the earth and, as some say, organic material (plants, animals, etc.) came into existence and made all of the coal, oil, gas and carbon and, since we know organic material has carbon, where did it get the carbon, to become what some say it made, all of the carbons of the earth? Does a son predate his own father?

It seems to be a well established fact that all fossilization terminates at the Cambrian yet we find billions of barrels of oil and trillions of cubic feet of gas deeper than Cambrian where there is no fossilization. Who put it there? One gas well in the United States of America produces endless amounts of gas from below the 27,000 foot depth. Russia now has either an oil or gas well drilling below the 32,000 foot depth. I predict that oil and gas, or both, will be found below the 50,000 foot depth when they can drill that deep; that there is a carbon ring all the way around the earth, even under all of the oceans, in various stratification, though not contiguous. All of this had to come into existence before fossilization. If everything is a combination of something else, where did the something else come from?

There are two men I try not to miss when I know that they will be on television: Dr. Paul Ehrlich, of Stanford University, and Dr. Carl Sagan, of Cornell University. Dr. Sagan was on the Johnny Carson Show on May 30, 1980. He has a tremendous knowledge of the physical operation of the solar system and universe. The entire universe is traveling through space, in unison, locked together by internal forces. When asked where we were going, he said, "Nowhere."

Dr. Sagan, in his very interesting book, "The Dragons of Eden," starts the first page with remarks from Charles Darwin. It seems Darwin, in his travels, came across some Fuegian barbarians (as he calls them). They were naked, had paint on them, long hair and a wild, startled, distrustful look; had no arts and lived off what they could catch. They had no government and were merciless to everyone not of their tribe. They delight in torturing their enemies, offer up bloody sacrifices, practice infanticide without remorse, treat their wives like slaves, have no decency and are superstitious. For these reasons Darwin said he would just as soon be a descendent of the heroic little monkey, who had better quality than these men.

Darwin's analysis is ridiculous, inconsistent and puerile. He claims that all men came from monkeys. Therefore, these men that he mentions must have also come from the monkey. Besides, his theory of natural selection as he claims is constantly upgrading man and his quality, morally and in every way. Darwin died in 1882. His theory has failed and if he were still here, he would see a world filled with men and deeds exactly as he describes; long hair and the works. In modern times, did not men come out of the Nazi camps and other man-made movements looking and acting just as his aboriginal man? Men who were normal and highly educated were reduced to animals, against his theory. The Fuegian men he here describes could be taken out of their setting and into civilized living and one day would be farming land, driving automobiles, flying jet planes and writing books. What monkey has ever done that or ever will? Darwin further says: *"Man, claiming he was not aboriginally placed, may give him hopes for a still higher destiny in the distant future. But we are not here concerned with hopes or fears, only with the truth as far as our reason allows us to discover it. Man still bears in his bodily frame the indelible stamp of his lowly origin."* Darwin does acknowledge that man has a Godlike intellect. Being created in the image of God is not a lowly origin. In His image does not mean men look like God, they just have a few of His qualities or intellect. Man, too, is a natural born creator or builder, as Darwin will testify.

Many evolutionists claim the universe always existed and never had a beginning. It is true that something or someone had to always exist, unthinkable as that may seem to us. Why then could not God, the Creator, be the one without beginning? Why is one position more difficult to accept than the other? Bertrand Russel said, *"A Creator is in need of a creator."* Why is this position any less provable than a creation with no beginning? Hebrews 3:4.

After many years of study, research and thought on the subject of the origin of life, matter, geology, archaeology, paleontology, including fossilization, I will give some of the reasons I do not believe in vegetarian geology, the theory that carbons, oil, gas and coal were brought about by decomposition of organic material, plants and animals. With as much free oxygen as we now have, they would not have become coal or oil but some other reduced substance or dissolve or decompose, etc.

My doubt of the authenticity of vegetarian geology began as a boy when my brother Gene and I were about eight and nine years old. We were learning in school, from textbooks, that trees and their leaves would accumulate under them until they would build up a thick layer of coal. We would go out in the trees on our farm where we lived and discuss this and wonder how that could be. The books said it would take thousands of years, yet we knew the trees would not live but a few years. We could see that each year the leaves fell off and decomposed into more soil; that there was no basis to claim trees would make thick, or even thin, coal beds. After thought and study we completely rejected the theory as false. There had to be another answer. As we grew older, we read the account of creation in the Bible which made more sense, although a lot more study was needed to decide this all important issue. God created all things we reasoned; therefore, all true science would agree with the Bible. Most of the information here is scientific, so there should be no problem.

THE EARTH — CARBON CONTENT

To arrive at the truth we must always go right to the very beginning of the source. Science today teaches that the earth at one time was a ball of molten hot material. This seems to be true; and that the earth must be billions of years old; that is true also. The six creative days mentioned in Genesis (not 24 hour days) do not apply to the earth's beginning but seem to be time periods used to prepare the earth for man to live on. We must agree then that the original igneous earth contained all of the elements we now have; it was a matter of separating them for useful purpose. A housewife makes a cake by mixing all of the ingredients together. How could we unmix them? Heat and centrifugal force, both already present in the earth, would do this unmixing.

Perhaps we should give some consideration to how much carbon the earth contains, coal, oil, gas, etc. Then ask ourselves if this could be only accumulation of grass, trees and animal bodies. Remember, the vegetarian theory was born when very little was known about how much carbon there was in the earth, or how many miles down it went, and that the deeper we go now, the more abundant it becomes beyond fossilization.

Starting with coal — it has been said that one state, as Wyoming, Montana, or North Dakota, has enough coal to last the entire world perhaps hundreds of years, not to mention other nearby states and all of the rest of the world. Much of it is at the very surface and millions of tons have been mined in the United States of America and North America by removing over-burden. Yet, coal of many kinds and quality have been mined as deep as beyond a mile. It is everywhere on the earth. When we drill for water, we find thin veins of it. It shows up inside hills or where highways have been dug through. A geologist, who is a friend of mine, said that in one oil well in this area they found a thick coal layer several hundred feet down. It has been found in a layer only one-fourth inch to one-half inch thick, over an area of thousands of square miles with no other coal in the area. Why did dead trees do this? Like oil, layer upon layer of coal in the same area is of a different kind of specific gravity, some hard, some soft, some lignite and anthracite. Why isn't it all the same if it has one common source, vegetation? There have been billions of tons of coal mined and many more left to mine, yet it may be estimated that there is just as much coal in the earth that cannot be mined as will ever be mined. It is too sparcely distributed throughout the earth for economic recovery. Almost every oil well will find some coal at various levels in parts of the earth and thousands of feet down. Are we to conclude that this was made by trees and grass? How did it get all through the earth in such huge amounts? We have not begun to discuss the subject of coal in the earth, but let's move on to an even more interesting subject, oil.

When we consider the amount of oil in the earth, we have even more problems in accounting for it by fossilization theory. When Edwin Drake drilled the first oil well on August 27, 1859, 69½ feet deep, a new age in energy was born. It was at Titusville, Pennsylvania, yet today, oil is still produced in that state and at much greater depth. This was not the first that man knew of oil as it is thought, the leaks in Noah's Ark were sealed with bitumen, as were Indian boats in northern Alberta, Canada. No one knows how much oil the earth contains. Only proven reserves can be guessed at. Many times it has been predicted we would run out of oil. Oil consumption in the United States has moved up to near twenty million barrels per day (42g) and has been dropping some since 1979. Perhaps just over 15 million in 1982. U.S. crude oil production in November 1981, 8,613,000 b/d. Same month 1980, 8,499,000. World daily use is set at near sixty million barrels per day. This, too, may be dropping some. Old cable rigs could reach about eight thousand feet, most not that deep. So they felt oil was running out. The invention of the Hughes drilling bit (by Howard Hughes' father) has drilling below 30,000 feet and finding oil and gas all the way down in various levels and strata-formations.

Silurian Age seems very good for oil. Are we to conclude that all this oil, gas, and bitumen, is nothing but dead animal life? Mere quantity rules this out plus its depth in the earth below fossilization. For years it was my feeling that as far as finding oil, that even the Williston Basin in North Dakota had not been scratched. This is proving to be right. Wells are scheduled below 15,000 feet now. There are 32 different geological zones in North Dakota where oil has been found; though not all in one well. Oil, like coal, is even more different than coal at all levels. Ranging from near 60 gravity to 14 in Wyoming, where we get a lot of road tar, then to bitumen in Alberta, Canada, and other tar pits over the world. It is more endless than coal it seems. In North Dakota we may be drilling where several layers of oil are found separated by a few hundred feet to several thousand feet and each with a different kind of oil as to gravity and other properties, and each will be different to the eye. The different oils will even refine differently and most refineries had to remodel to handle Alaskan crude oil. There is sweet and sour, etc., high and low sulfur, etc. Some wells have a very high salt water production, others less. Some wells make far more water than oil. Why so many kinds of oil all over the earth if it is just a combination of dead animal life?

Much geological information has been helpful yet most all predictions of where oil had to be found by most early geologists proved to be wrong and even in recent times, is wrong. As one world famous geologist told me, oil is where you find it. There is just too much oil to be a combination of something else, it had to be a direct creation during the millions of years of the carbon age, before plant or animal life.

There is an energy crisis, be sure of that. My point is that we have not used up the carbon energy; the demand is just greater than man can produce. If he did, we may all die of carbon poisoning of the air. If the gas shortage was contrived, higher prices cause drillers to spend more to drill deeper to find greater amounts of gas and oil. The Tuscaloosa Sand gas find in Louisiana (some at 21,345 feet and some now much deeper) has tremendous amounts of gas. According to a Time magazine (December, 1977, issue), one well produced gas at the uncontrolled rate of 140 million cubic feet per day and now will be controlled at 20 million cubic feet to supply 61,000 homes. It would seem to me that the deeper oil bearing strata of the earth has more oil and gas than all of the known reserves. The World Petroleum Congress, meeting in Bucharest in 1979, said the Arctic and the depths of the oceans hold vast, untapped oil and gas reserves. It would seem that there is more oil and gas above the Arctic Circle than in all of our known reserves today.

One of my favorite geologists and writers, the originator of the theory of the earth's annular system, Isaac N. Vail, who died back in 1912, told in his book, The Earth's Annular System, where most of the great oil fields of the world would be found, before they were even dreamed of. They are right where he said. He proved vegetarian geology was wrong. He also told where oil would not be found and it still has not been found there. He said one of the best prospects for a big oil find would be in the North Sea, as well as Alaska, etc. It has been said now with wells making up to 80,000 barrels per day, some even more, it is equal to the Middle Eastern Oil Fields. Also Siberia, and some feel Alaska has more gas than Texas. Mexico and the Gulf is now thought to contain unlimited oil and gas. I feel the trend goes all the way into Central America.

There are still several large amounts of carbon to account for, shale, and heavy oil sands, as in Alberta, Canada. Experts now say, and they seem to know, the Athabasca oil sands contain billions and billions of barrels of oil. More than the amount in all of Saudi Arabia and the East. One place they set at 800 billion barrels. Another at 600 billion. Another at 150 billion barrels. The list goes on. Then what about the shale mountains of Wyoming, Utah, and Colorado, some say equal all known oil reserves. My point is still, where did all of this oil and carbon come from, how did it get there even up in mountains? Will fossilization answer this? No way! There is still an endless carbon amount to mention next.

We have discussed oil, gas, coal, from the surface of the earth down to 30,000 feet and as I said I feel it will be found to 50,000 as many others do. Proven amounts of them are beyond human comprehension. Now let's talk about an amount no one ever mentions, the unrecoverable of these products in the earth. This, it would seem, is an amount equal at least to the recoverable or greater. To explain, an oil or gas field is found within the boundary of the rock, sand, or shale, whatever it is in. Oil must have an impermeable barrier on top of it and all around it or it would have drifted uniformly throughout the earth's strata with no trap or pressure to make it recoverable to surface through a pipe. The material on top of oil must have a low porosity impermeable to oil or gas. When a drill bit goes through this and pipe set, it can flow up under pressure or be pumped up. Some gas wells I have seen have a 10,000 psi. wellhead pressure, some may be greater. There are no open pools of oil or gas in the earth, as some may believe, it is always within strata.

What I mean by unrecoverable oil and gas is, just as there is unrecoverable coal, too thinly distributed and too small amount yet millions of these places all over the earth. Many times we hear of someone drilling for water and getting gas. We know of people who have been able to heat a garage on a farm, etc., from a water well making gas. Yes, the earth is filled with carbon in all forms which did not get behind a porosity zone and just drifted until it found a neutral position. Some gas and oil, especially years ago, did find its way slowly to surface and escape into the atmosphere. Fires in farm fields would sometimes have a ribbon of blue flame for hours in the cracks of the ground as some gas was escaping before it was used up. Yes, unrecoverable carbons in the earth are never part of reserve figures, I only mention them as we are trying to establish the total amount in the earth and how it got there, and this does not add any substance to vegetarian geology. All forms of life are carbon and cannot have created what they are, without carbon they could not live in the first place. All forms of life are parasitical, one form lives off another form. In the ocean the whale and walrus are the end of the food chain.

I am getting away from my subject, let us say we are drilling an oil well in western North Dakota on the Nesson Anticline and go through 5 sections, or layers, of oil producing strata, each with an impermeable cap rock on top of it, each with a different kind of oil, each separated by a few hundred feet to several thousand feet, each with a different bottom hole pressure ranging from several hundred psi to perhaps several thousand psi. Could there ever have been any vertical migration of fluid here under these circumstances? Did this oil just run down into the earth as some would say? It does not seem possible, does it? The oil and gas and coal had to be laid in at the same time the fill in was, or nearly then, and it has changed very little from the time it became static until now. Some oil producing sections are only 2 to 3 feet thick, where some good, big wells come from a section many feet, even in the hundreds, and some like the Middle East in the thousand feet. Not all oil bearing structure is the same, layer upon layer, even in North Dakota. Some is very porous, others are tight, and oil moves slow. Some places on earth, like Arabia, the oil comes from sand formation, etc., and a well can make thousands of barrels of oil per day with no damage to the structure.

One well I read about in the North Sea makes 80,000 barrels per day. Like flying an airplane, no two landings are ever the same, nor are any two oil wells ever the same even in the same oil field. This rules out vegetarian geology. Some wells in North Dakota have a bottom hole pressure of 4,000 pounds and some even more. My point is, fluid will not run down through pressure like that, it would rather move up to neutral area. Vertical fluid movement would not be possible anyway with impermeable barriers, cap rock, on all oil reservoirs.

This is not to say there could not be some minor vertical oil fluid movement in oil zones which are similar and lie close together, where fractures may exist. These kinds of zones would not be deep zones or separated by thousands of feet. It is in total concept throughout geological times after reservoirs were permanently fixed, we say there is little or no vertical fluid movement.

Some experts feel deep basin reservoirs of gas, down to 30,000 feet deep in the Gulf Coast and west, will contain 215 trillion cubic feet of gas. Others feel Devonian shales laid down in the Northeast have up to 600 trillion cubic feet of gas, though these dense tight shales have low porosity, and produce slow. Western tight sandstone holds 50 to 600 trillion cubic feet of gas. These can be fractured though, which may free some gas. We are referring to thousands of square miles of the earth. Where are the marine forms they say formed all of this so evenly and tightly locked in the earth? Fossils are more numerous near the surface and get fewer as we go deeper down and disappear at the Cambrian Age. Yet oil and carbon increase. North Sea wells are as far as 200 miles offshore. Was the ocean always over this oil field? How did the oil get there in such large amounts?

OIL RESERVES

As with shale, tar sands, and gas, there is also a vast amount of unrecoverable oil left in the earth, an amount perhaps greater than all of the recoverable and known oil reserves of the earth. This is oil drifting, or now at neutral positions throughout the entire earth, that did not find a porosity barrier, impermeable material to lock it in. We could drill right through it and the cutting samples may have a showing, but none of it will ever be recoverable. This I would also say extends all of the way around the earth, even under all of the oceans and continents, though not contiguous, and in many stratifications throughout the earth's 50,000 feet of crust. This amount must be greater than all recoverable and known oil reserves today. Where did it all come from? Can it be organic matter which, for the most part, decomposes with oxygen present? Doesn't seem possible, does it? Some thought on this subject is that oil and gas are developing or generating continuously in the earth. I do not see any evidence at this time to support such a claim. Such a metamorphic action just does not seem possible. Where would the fuel come from to cause such an action?

On the subject of oil's origin here is a theory I have supported and talked about for many years. This was published by L. Gaucher in Chemical Technology: *"I suggest that oil could have been formed, in much larger quantities than was ever considered plausible before . . . long before there was any plant or animal life at all . . . Instead of assuming that oil was formed under surface and atmospheric conditions similar to those that we find on earth today, as the organic theory does, I suggest that oil was formed through chemical reactions of components of the atmosphere at the time when the earth was still hot and devoid of life."* Nineteen hundred seventy-three, Goucher, a chemical engineer, suggests that an "Oil Rain" brought this mixture to earth. This idea seems far more reasonable than the organic vegetation theory.

At this time I will bring your attention to an article in the Wall Street Journal of June 20, 1980. The article on page 8, is by Mr. P. W. J. Wood, President, Energy Resources Group, Cities Service Company. If you can get a copy of this paper read the entire article, it is very good. Mr. Wood says, *"Experts have been predicting since 1860 that the world is running out of oil very rapidly. World oil reserves have long passed the one trillion barrel mark. Modern forecasts indicate at least one trillion more, perhaps another three trillion. Forecasts have been conservative in the past because they are tied to economics and technology. He mentions many giant discoveries like the onshore-offshore of Mexico, with probable reserves of 200 billion barrels as well as gas. Also the Overthrust Belt of the United States. Technological advances now make it possible to drill much deeper and in places never before reachable, like iceberg-infested environments and arctic areas. Gas is now produced in the United States at depths exceeding five miles. If there is a widespread petroleum shortage before the end of this century, it will not be because reserves are not available to be found or because industry lacks the technology. Political factors, rather than technical ones, will determine whether enough petroleum will be found to permit an orderly transition to the other forms of energy in the 21st century."*

Mr. Wood gives a very optimistic report on oil reserves, to which I would agree. Not being a vegetarian geologist, I would be willing to increase even his expert figures. Over 25 years ago I believed the great oil reserves of Mexico, the North Sea, Alaska and many other places were there by research in annular theory. One author placed them there nearly 100 years ago, not by vegetarian geology though. I believe oil and gas reserves of the earth are boundless, beyond comprehension, yet to be discovered. Who will have the money and know how?

I am having trouble finding words to reintroduce the subject of carbon in this edition of my book. We must start with a known and universally admitted premise. We must have a boundary to start from. We need to reduce something to fact and truth. Physical law is universal. Carbon is universal, as that has been pretty well established now with the Viking, Voyager, and other space missions. Mars has carbon dioxide, and some oxygen. Down here on mundane earth, they say all carbons are made of organic plant and animal life. What then made carbon on Mars and other planets, and their moons? It is a well known fact there are no plants or animals, or any form of life on them, from which to make carbon. Carbons range from pyrophoric (coal) to nonflammable (gas). How can we fit the puny vegetarian single process into this complexity? Toward the front of this book I made the comment — there is more oil and gas north of the Arctic Circle than all of the now known oil reserves of the world. I mention wells at Hibernia and the North Sea as evidence. Experts agree, western Siberia does contain large quantities of crude oil, perhaps the light sweet kind. As is true of much oil around the globe, this Siberian oil is trapped in a tight rock formation difficult to extract. The news announcement of December 5, 1980, of the "biggest" oil find in history in western Siberia may very well be an exaggeration, even though a large oil deposit exists.

Let us ask the petroleum engineer, the microbiologist, molecular-biologist, and all scientists, if, in their frank, honest, expert opinion oil and gas are made up of animal substance, considering such are organic in nature, while hydrocarbons are inorganic? In such a transition, how do we account for enzymes, amino acids, and a host of other elements in organic animal life, not found in the composition of oil and gas. If this were actually the case, would not the components of oil and gas be different? It has been said, about 700-1,000 products and compounds may be made out of carbon, by the catalytic process. Let's reverse the process for an answer, which is what we must do anyway. If dead animal material is the granddaddy of oil and gas, how many products could be made out of this residual of life? I am sure you see what I am getting at. Never can we make more out of less.

If, as we have been taught, all carbons are of a vegetative derivation, what mystic process placed carbons on the other planets of the solar system, and maybe, even beyond that boundary? Our "premise" must be, carbon is a universal product, distilled in the igneous beds of the planetary systems, of the solar system, or beyond when proven. Carbon is so great and overpowering in amount and structure, it cannot be tied to the puny vegetarian process. It is evident that every world with an annular system (rings) must have a supply of unconsumed primitive carbon, as one of its prominent components. By ransacking the great laboratory of nature, could we find any other element that can be made to take its place?

Carl Sagan on his TV show, Cosmos, said, *"It took billions of years for the evolution process to make one bacteria, or grasshopper."* Can you believe such an assumption? He also said, *"When we have strong emotions, we may fool ourselves."* It seems that is what he is doing. He said, *"The study of Mars is deeply ambiguous, liable to be fooled on life being there. Life is chemistry, maybe we will find life on Mars not made of organic chemistry."* Microbiologists and scientists feel microbes will be found on Mars, and other planets. It seems easy to make such claims, then if life is found, they may say: I told you so, and take the credit, even though the odds are totally against it. Sagan feels life could be planted on Mars, and other planets, and we will become the Martians. First, we must develop some form of plant by genetic engineering, and transplant it on Mars. Plants placed there would first use carbon dioxide, and produce oxygen, in time Mars would be livable by humans. It does seem ironical, this is the exact process the earth developed, without the manipulation of man. Yet, who is to say his scheme is not in the Great Plan of things? A large farmer must start somewhere on his land, when the spring planting season starts. If the universe is to have life in other areas, why not start with the earth? Just a thought!

Dr. Sagan, on one of his shows, walked into a no-man's land of speculation, by saying a small fish in ancient seas, over millions of years, developed lungs for breathing oxygen. This small fish, he claims, was the forerunner of all life on earth. Can fact and reason account for this long gray area of time, from nonbreathing to breathing creatures? In this unbelievable, and unprovable transition, how do we account for the fact that oxygen is the life sustaining force of man and animal, and none can live but minutes without it? Who would dare to suggest otherwise, or attempt to prove it was ever different?

CHAPTER 2

FOSSIL EVIDENCE?

Page 114 of Moody's Fossil World says, *"Fossil evidence remains a conclusive testimony in support of the concept of evolution. The documentation of horse evolution alone provides ample support for this statement."* This is not a correct statement, as no one has ever found or proven any inbetween kinds of any life form. They appear suddenly on the scene, and if they disappear, they leave just as they were found. Yes, the study of the fossil world is very fascinating and deeply interesting. It has been said there are 500,000 kinds of insects logged, and 500,000 left to be logged. Yet, there just are not enough fossils to make all of the earth's carbon. No way at all.

There is an interesting observation on page 283 and 322 of Earth's Annular System. Quote: *"We find vegetable remains in coal seams just as we find them in other rock. It is easily demonstrated that the vegetable carbon in the coal beds is generally not bituminous even in a bituminous bed. It is just carbonized vegetable debris, and it is scarcely combustible. Fossil plants in coal are generally mineralized charcoal, and difficult of combustion. If the bed were bodily a vegetable product, the same difficulty would certainly characterize the mass and we are, therefore, compelled to admit that the plant is simply a foreign body in a bed of mineral carbon and itself a mineralized carbon fossil simply because it is in that bed. In short, we are forced to look beyond that plant for the origin of the bed."* We cannot consistently say the range of coal is caused by metamorphism; no, it was always just as it is since it was laid in place through creative processes. More evidence against vegetarian geology. Time Magazine for March 31, 1980, has an ad AMOCO Oil Company, saying they have a 5,000 acre tract of shale in Colorado, that could produce up to 5 billion barrels of oil. This is only a fraction of shale oil on earth. How did vegetarian geology fill this mountain with oil?

Let us just run over the carbon of the earth again, even though no human could ever even come close to an accurate estimate of how much the earth contained and still has left. Some countries of the earth flare in the atmosphere more gas per day than the entire consumption of the United States, just to get the fluid condensates, etc., which are easy to handle and a ready market for. Because of economic reasons, Mexico, and to even a greater extent, the Middle East, flare millions of cubic feet of gas per day. Consider first the amount of gas, oil, and coal the earth has already given up. Then think of the vast reserves left yet to be produced; then think of the vast amounts not found as yet; then consider the shale and tar sands; then add the unrecoverable amounts of all carbon, which are perhaps far greater than the recoverable. Are we to conclude this all came from a few fossils, which are hard to find even in the upper strata, and completely absent in the deeper strata where I say, as others do, our greater oil and gas reserves will yet be found. It does not seem reasonable, does it? Then there must be some other answer. In other words, fossil fuels, as they are sometimes called, are not fossil at all.

In his book, the Fossil World, by Richard Moody, he stated on page 117, *"Oil and natural gas are probably the result of the accumulation of silts with high organic content. And, if trapped within would undergo decomposition, in the absence of oxygen, to produce the various hydrocarbons."*

If this is true, it would only produce one kind of carbon. Then he says, *"The generation of oil from organic matter is still poorly understood, but the investigations of paleontologists and petroleum geologists include the study of both source and reservoir rocks."* On page 116 he says coal, oil and natural gas are the result of decaying organic remains. He then says coal ranges from lignite to anthracite, with sub or intermediate ranks as bituminous. If this accepted process has no variation in it, why would it produce so many kinds of carbons, and hydrocarbons? Also, we must account for oxygen or the lack of oxygen. Who controlled the valve for the oxygen? His book is very interesting and educational, yet his premise is wrong and not consistent. The study of the fossil world is indeed interesting and fascinating, though I see little evidence fossils could produce the vast and varied amounts of carbon we have accounted for. It is with great interest that his work starts at the Cambrian Age and moves up to current, or surface. Yet carbon exists far below Cambrian (deeper, if not under).

SALT

Another geological mystery not answered by vegetarian geology is salt content of the earth. In no way can I discuss all of these in detail; but, for example, in drilling oil wells in North Dakota, in places we find large salt deposits. This may be true in many places on earth. Sometimes we find a layer of salt from several feet thick to several hundred feet thick. There is one oil well in North Dakota changed to a salt well. These salt layers cover many square miles and are pure salt. Salt is impermeable to oil and gas. Yet much oil and gas is below the salt layers. How did it get under there if it is impermeable? It had to be laid in there first, I would say. How can we explain the Piercement Salt Domes in the Texas, Louisiana, Gulf Coast area? Some even in the ocean water. A circular dome of salt, seemingly pierced up from below, coming up within a few hundred feet of the earth's surface, several miles across, which will have many oil wells drilled around the dome down the slope sides, some of which have made oil since 1938 at the rate of a couple hundred barrels per day and no one seems to know where the oil comes from. It just feeds up the impermeable side of the salt dome from a load deep in the earth, no one knows where from. Reservoirs cannot be accounted for, so wells do not have spacing control. Just drill as many as you want as long as they are not illegally slant drilled. To my knowledge, no one has ever drilled to find the bottom of a major salt dome, it is just too deep, salt all of the way. They do flush out the salt and make large oil or gas reservoirs though, to store oil products in. The point is, how could vegetarian geology cause all of this amount of oil to gather deep down at the bottom edge of the salt domes? It does not seem possible.

There had to be a time when the salt content of the earth was fixed in total amount. Some have claimed there is evidence of less salt content or a change in salt content in some upper geological things and features in the earth. They seem to feel this helps prove an ice age. The reverse is true and it supports my theme of annular theory. It is my claim there has been tremendous ocean augmentation by water in late geological times. We know the salt content of the earth had already been permanently fixed, so the addition of a large amount of salt free water would cause a dilution of the salt content near the earth's surface, without ever reducing the total amount. Again, the answer is very simple, it is just a dilution factor, nothing else. (See salt dome picture in my book page 229.)

In the interesting study of the origin of life on earth, and the origin of the solar system, as well as the universe, modern science leaves too many questions unanswered, and each new discovery now disproves everything they had learned before and accepted as truth. There are so many subjects to cover I will touch on them only a little, yet try to make a point. For too long we have accepted theory as fact, most of which is not provable, as if it were the very Word of God, just because it comes from some recognized, so-called source of authority. As many other qualified persons do not espouse nor accept many ideas and theories on the above subjects, I do not either. The ice age or glacier age seems to be a myth only, and unprovable. What force would move a large amount of ice to cover the North American Continent? It seems water shaped our earth as we know it now, not ice. Many of us reject the idea that great dinosaurs died and made the oil fields of the earth, even though it is a proven fact they once lived, perhaps before man, or until the flood. For hundreds of years, millions of buffalo roamed the western United States and died or were killed by man; yet, do you believe they made one drop of oil in western wells? For years some have been trying to prove man has been on earth for up to 3.5 million years, yet there is absolutely no evidence man has been here that many years. Prehistoric means before history; therefore, it is guess work, or just made up as they go along without proof.

CARBON 14 DATING

The main problem with Darwin's theory of evolution and the survival of the fittest is the opposite is true. The weak destroy the strong in most forms of life, economic, social or literally. They always encounter the strong, as in the virus world. Most life forms have gone backward, including man.

Carbon 14 dating, once thought to hold answers to the age of everything, was called a "Real Bag of Worms" by Nobel Prize winner Willard F. Libby, Popular Science, June 1971. We all remember the Piltdown Man was nothing but a hoax. How many others are just a hoax, or just made up stories? A number of scientists wrote in the Popular Science Magazine in November, 1979, on Carbon 14 dating mystery. Physicist Robert Gentry, of Columbia Union College, believes that all of the dates determined by radioactive decay may be off — not only by a few years, but by order of magnitude. As he says, *"Man, instead of having walked the earth for 3.6 million years, may have been around for only a few thousands. Further, that the rates of decay are forever unvarying — an untestable assumption."* Popular

Science Magazine, November 1979, page 81.

Another article in Popular Science, April, 1980: Meteorites, Key to Astral Secrets, does not prove evolution theory, or origin of Solar System. Many meteorites are not made up of the same material all of the way through. Some even look like a loaf of bread with raisins in it. No one seems to know where they came from. Regardless, nothing so far would disprove the Genesis account.

This is a Triceratops dinosaur fossil found near Marmarth, North Dakota, in 1965, on display at the Lobby Museum, Leonard Hall, University of North Dakota, at Grand Forks. Many kinds of dinosaurs are found in the central U.S.A., most on or near the surface. Therefore, they could not have been lying there for many millions of years nor could they have made any oil found miles deep in this area.

Many unprovable claims are being made, as an article in the Los Angeles Times, dated February 25, 1978: Prints of Man's Ancestor Believed Found. Footprints 3.5 million years old said found by Mary Leakey, widow of Dr. Louis Leakey. She believed she was 75% certain of these prints found in Tanzania, Africa. The Bible and history will show man's beginning was Mesopotamia, and not Africa. Who could believe such stuff anyway? I am 100% sure these footprints are not anywhere near that old regardless of what or who made them. How would anyone be able to put such a date on them? Could a footprint last that long in strata? Fargo Forum Article, November 19, 1979 — 1.5 million-year-old footprints found. Who decided how old it was, is my question, and where is the proof? We all recall the so-called Big Foot in this area.

A recent copy of Time Magazine has a picture of a relic from North Dakota. It says, *"Triceratops found near Marmarth, North Dakota, in 1965, on display in Lobby Museum, Leonard Hall, University of North Dakota, Grand Forks."* It does not say if they believe there is some connection between this fossil and oil found in North Dakota; yet, may I remind again, some oil now found in North Dakota is below 14,000 feet, such as the Governor Link well, and I understand some will drill below 15,000 feet soon if not already. Bone fossils in North Dakota are all found near the surface, how did the oil get way down where it is if these fossils contributed anything to it? The Fargo Forum article of August 27, 1978, about crocodiles being found in the North Dakota Badlands is very interesting. Paleontologists say this area was once tropical, perhaps like Florida. Some put a date of 100 million years on such fossils and bones. Yes, it is true North Dakota as well as Siberia, and Alaska were once warmer or even tropical, as many bones and fossils are found there, too. Also, oil drillers on the North Slope find evidence of tropical life when the bit goes down into the tundra. Certainly the earth had a different climate than now. Many animals in Siberia died so suddenly they still have grass in their mouth, and the meat is still used for food when found. What happened to do this?

A GREAT DOWN-FALL OF WATER

A great down-fall of water would be the only answer to such a great cataclysmic destruction, which caused the earth's climate to change and its surface to be so shaped as it is today. This sudden destruction could also allow for the freezing of the animal bodies as in Siberia and other places leaving them intact, buried under earth and ice for centuries and being dug up now. I have one book on geology which says, *"It took the Colorado River millions of years to cut out the Grand Canyon."* In many visits to the Grand Canyon, I do not see any such evidence in time or otherwise. The bottom of the deep part of the Grand Canyon they say is the Cambrian, so we see a cross section of geology up to the surface. Also, the Canyon is 5,000 feet deep. May I ask why water ran uphill for millions of years until it cut the Canyon, while on either side it was stratifying the geological ages they mention, with a sprinkling of fossilization, here and there?

This same book mentions the Badlands of South Dakota as being formed over millions of years. The South Dakota School of Mines put the Badlands there at many millions of years, yet the North Dakota University calls the North Dakota Badlands millions of years older. I do not see any difference in the age of the Grand Canyon. It would seem they were all caused by the same thing and at the same time. Not a mythical ice age, but a great deluge of water. This is our clue to what great forces it took to cause earthquakes, heave up mountains to much higher than they originally were, bend and fold bedrock. The earthquakes of today are just a result of this action as the earth is still settling toward a neutral position, the natural position of all force.

The flood also answers the question as to claims of continental drift. Even though there may be some small amount of continental drift or shifting of the earth's crust even today, yet the addition of more water from outerspace to the earth would press all ocean floors down some and heave up mountains higher and raise all shore lines answering the questions for so-called continental drift. Continents separating by thousands of miles just did not happen.

No one, as yet, has been able to prove how the original earth and solar system came into being. Most will agree the earth was very igneous, and perhaps too much importance is now given to igneous and sedimentary rock still forming into metamorphic rock. The greater part of this age is behind us now, or we would not be here, nor could we live if it were not. Too, many questions are left unanswered by vegetarian geology. My geological book says glaciers in Alaska took millions of years to form and do their work, yet why is it, right under them tropical plant life was once in evidence, and millions of animal bones are dug up by dozers when leveling ground for airports, none of which live there now, and do not seem to be very old, perhaps a few thousand years. If sedimentation has always been constant as some claim, why do they find fossils and other material, like footprints, as some say near the surface? It would seem to me, things are pretty much as they were put, layer upon layer with the only surface change taking place at the Flood. When they claim the markings or fossils are millions of years old, my point was, should have disappeared, by their erosion claims.

CHAPTER 3
THE HISTORICAL FLOOD

Perhaps now is the time to discuss a little on where the flood water came from. We know such a great fall of water would not come from ordinary weather patterns, and evaporation processes as we know them to be. The earth is 29% land and 71% water surface now. Before the Flood there would have been more land area and much less water on the earth. At present if the earth were all level and no mountains or valleys, etc., like a ball there would be 8,000 feet of water everywhere on earth. Where could so much water come from? Some have suggested there was once a great water canopy or belt around the earth far out into space, causing a warm climate all over the earth, accounting for the tropical plants and other life, living in now cold parts of the earth. The flood saw this belt of water come down to the earth, causing it to shape as we know it now. However, for those who do not believe this, nothing will change, things being as they are, regardless of what different individuals may believe, just will not change that. Man's physical environment in antediluvian times was just different from now. The sun would come through the water ring, it did not rain on the earth as we know rain, from the creation of man until the flood. That is why the words at Genesis 8:22 were not written, or apply, until after the flood, quote: *"For all the days the earth continues, seed sowing and harvest, and cold and heat, and summer and winter, and day and night, will never cease."* Before the flood such seasons did not exist, but now will forever. All evidence in geology seems to indicate the shape of the earth, as we know it today, was caused by water, not so-called ice age. Plant life, or other forms of life are not associated with an ice age millions of years old, as the claim is made they are.

It would seem as Vail says, *"There is a natural scheme found prominent in all earth-building that harmonizes into the establishment of a law of progression which nothing short of the decline of rings and the reign of earth-canopies successively can satisfactorily explain. The 'Old School Geologist' will likely be the last person to admit that the earth's aqueous strata have largely fallen into place as we find them today, as the giant wreck of slowly declining annular matter, that an annular system does not necessarily fall in as a sudden titanic world collision, but is continuous over a very long period of time.''* The original igneous earth would cast off much of its material into space and as it cooled and slowed some, it would all come back down again layer upon layer, just as we find it today. The flood waters being the last to fall in. The recent explosion of the volcanic Mt. St. Helens, in Washington state, is a very small example of what is meant by stratification build up, by an annular system. All of the ash will eventually fall back to earth somewhere. Imagine the entire earth igneous, add centrifugal force, plus other factors and we have a world building process, given millions of years time.

The mind is utterly at a loss to find figures wherewith to multiply the vegetative process, to account for all of the carbon in the earth in great amounts accounted for in the above material. Only a primitive world-furnace could accomplish the carbon creation in such unheard of amounts. Yes, the primitive igneous process could have distilled all of the carbons of the earth long before plant life ever existed. I will again quote from Moody's Fossil World, page 116: *"Fossil fuels, such as coal, oil and natural gas, are the result of the accumulation, and decay, of organic remains."* Also he says, *"The rank of coal is related to the degree of metamorphism that has taken place since accumulation began. Coal ranges from lignite to anthracite, with sub-bituminous and intermediate ranks."* There is no way this theory can account for the billions of tons of many kinds of coal located all through the earth's crust. It would seem that this is not true at all, nor can be proven. It would seem the earth's strata today is just about exactly as it was when it was placed where it is found. All too much emphasis is placed on metamorphism. Most matter has changed very little in current times. Moody says, *"In many ways the theory of continental drift and plate tectonics has been the cornerstone of modern geology, with paleontologists, sedimentologists, structural geologists and geophysicists working together towards a common goal."* This is not correct, as they disagree completely at times among themselves, each trying to get himself the publicity by making some far out unprovable claim.

As I am writing this a new oil discovery proving billions of barrels of new reserves was discovered. According to the paper of March 28, 1980, the well found in the Atlantic, off Newfoundland, at Hibernia, was a well about 8,000 feet deep with a sand pay section 250-275 feet thick the report says. It has the potential of being 100 billion barrels. It was also said more Hibernia's will be found. Where did all of the dead marine life, etc., come from to make this oil? To quote the Wall St. Journal: *"Five years ago, if a geologist recommended that a wildcat be drilled in the western overthrust belt, he would have been considered stupid.''* Now they estimate at least 20 trillion cubic feet of gas is there, plus oil, etc. Also reports in both Time and Newsweek, since 1977, giant gas gushers have been found in the Tuscaloosa Sand in Louisiana, as well as Texas, and Oklahoma. One expert said two such wells could produce the energy of a number of nuclear plants. Most of the best geologists in the world did not recommend drilling in these places just a short time ago. Yet the great schools and university institutions expect us to believe every word they say on the origin of life, and all matter on earth. They have not been right on many subjects before, so why should we put faith in them now? I will add one more news item just in of a major natural gas discovery in the North Sea, by Dutch-Shell Group. Government and industry reports say the gas field could be the largest in the North Sea, and among the largest in the world. Reserve estimates are placed as high as a possible 56 trillion cubic feet of natural gas. See U.S. News World Report, July 14, 1980. Subject of Synfuels.

METAMORPHISM OR BLENDING

An interesting article in U.S. News World Report, November 25, 1974, LaJolla, California. *"Deep Sea Drilling — Scripps Institute of Oceanography. What was once a chain of lush tropical islands stretching for 1,500 miles off the Coast of India can now be found a mile below the ocean surface. The islands sank, they say."* True, these islands sank as did many other islands now found under the oceans of the earth. What other known great force could do such a thing as the Flood; which also increased the total amount of water on the earth moving up shorelines as well as covering some areas not covered with water before. We do know mountains seem piercement in appearance by the slant strata, as Mount Rushmore seems to be perpendicular at the top and slanted up from all sides, until the very top. Are we to argue because ants build small hills, or mounds, that given enough time they made all of the mountains? Or, perhaps, because a few fossils may be found in a coal bed, and yet many more are usually found in the clay just above the coal bed, the fossils also made the clay? I am sure no one would claim this, as fossils are found in many kinds of earth and rock. Who will decide which they made and which not? It took an annular system, to float in all of the world's debris, hundreds of thousands of years to do this monumental task. This Ring system must have started long before organic life began and lasted until such life appeared, accounting for the fossils in some carbon as well as earth strata.

The word metamorphic, or metamorphism, is used very often in geology books. I would say it is overused, and is just an easy answer for a theory which does not seem to have an answer, from their point of view that is. It seems to be an excuse for unprovable claims, or little understood, it makes a good cover. Consider just one element, say gold. How can constant erosion and sedimentation account for its distribution throughout the earth? Some has been found in thick and thin veins and mined from the earth near the surface. Still other gold is found in deep mines. Much gold has been found in a very small trace as deep as 8,000 feet, like the Homestake Mine, of Lead, South Dakota. It takes a ton of material to get just a few ounces of gold there. Yes, and even silver and other valuable minerals are found with it as a by-product. It would seem the annular ring system, over a long period of time, could accomplish this uneven task; but how could simple sedimentation, and constancy of erosion, do such a job as claimed by orthodox geology. We do not have a uniform geologic boundary to work from. Why then, would uniform erosion do such a varied job of sorting all other minerals down through geologic strata? I do not mean to imply there is no such thing as metamorphism in geologic development. First, we must have an active agency, which is mainly igneous. The igneous age for surface geology is past, before, or at final placement. (Exclude radioactive decay.) This subject will be discussed more later, even though I am not always keeping the exact order I would like, but will move along to another subject.

GRAND CANYON — EROSION

It has been my pleasure to go to the Grand Canyon a number of times over the years. In 1979 I spent several beautiful days there with my wife and daughter, Nancy. I think I drove them up the wall, or perhaps the Canyon, with my constant geological comments about this and that. As I stood on the Canyon rim, I saw no evidence of the Colorado River running uphill for 40, 50, or more millions of years to cut the, now, river valley nearly one mile below. The only possible answer I have ever been able to come up with is the Canyon was caused by the Great Flood. Geologic times since then seem to me to be quite static, except for a very small erosion factor and the silting of the river, or rivers, not much has changed. Standing on the rim we see tall, towering pillars of rock, long, slender, vertical, appearing ready to fall over at any moment, or a large boulder precariously on the rim edge or on top of one of these pillars, or perhaps a natural arch like Angel Arch, perhaps just as they are now, for centuries. As we stand in the stony, relentless silence, our perceptive mind will tell us a great amount of water must have fallen, and washed the softer soil and earth all away, and left the rock and harder material just as we see now. We could say the same about the Badlands of North Dakota South Dakota, Montana and everywhere on earth. The same color scheme is found in the Arizona Painted Desert as the Badlands. Perhaps we should assign them the same age, and same cause.

It is, therefore, easy to say and prove that before the flood there was little or no erosion from wind or water, or river silting. Wind currents, as we know them, and rainfall, as we have now, did not exist then. Nor were there any large rivers flowing, as without heavy rainfall they could not drain. Mist and atmospheric moisture was all that was needed to keep life's ecosystem sustained. Let us take the geologists and scientists own erosion and silting figures and apply them to the millions of years they claim it took to develop large rivers like the Amazon, or Colorado. If the silt figure of 500,000 tons per day is correct for the Colorado River, and/or other rivers on earth, for all of these millions of years, would not the erosion and silting process have run its course thousands of years ago, and there would be nothing left to silt or erode? The process would now be neutral, the ultimate end of all force. So, unless the earth would bulge or distort, there would not be water drainage at all, only float from rainfall. We must conclude the earth's stratifying process ended long ago; and, erosion by wind, water, and river silt is not very old, at most a few thousand years. There just is not enough material to fulfill the figures and time the scientists allow the process. From a scientist's point of view we must combine all erosion factors, wind, water (by rainfall) and silting.

Perhaps it is true, at times there are both erosion and sedimentation working together at the same time. Certainly we have both today. Yet, if they are, we must deal with the law of displacement and negation. This would raise havoc with the theory and the claims of most textbook geology. I have a book by Time-Life on the Grand Canyon. This is a very interesting book and has many wonderful pictures, and even though I do not agree with the age they give the Grand Canyon or the way it came about, it is still very educational to read with many facts in it. I am not against books, we should read them all to learn. On page 70 the book says, *"Many millions of years ago the Colorado Plateau in the Grand Canyon area contained 10,000 feet more rock than it does today, and was relatively level. Approximately 65 million years ago, the plateau's flat surface in the Grand Canyon area bulged upward."* May I ask, where did this mass of rock go to and what force caused it to leave? This explanation given seems to be an effort to show regional elevation much higher to account for drainage to create the Canyon, as I said earlier, water could not run uphill for millions of years to cut such a canyon.

Every evidence is that the Grand Canyon, along with many other features of the earth's surface geology, was caused cataclysmically, not over billions of years of erosion and sedimentation, even though we agree the earth and its original rock, etc., could be billions of years old. The book does accurately say and estimate the average daily silt of the Colorado River is about 300,000 tons, flood stage more. Dams such as Glen Canyon and Hoover, now catch the silt and sand behind them. As they say, the reservoir cannot forever hold this endless sand and silt and will leave a fearful problem for some other generation to decide what to do with inland seas of mud behind useless dams. The slanted wedges of rock remaining bear witness to a great flood of water, not what the book claims. These wedges are piercement caused by sudden upheaval from billions of tons of water from outer space falling in. Legend has it from Indians still living in the Grand Canyon, it was made by a great flood of water. Page 99. A very interesting book.

I do not mean for this to be an attack on every endeavor of man. Not at all. I consider myself a very scientific person. We do have a right to question though, as little children use this as a great learning tool. Let's have some good answers then. Many things held sacred by men are just not. We are often intimidated into acceptance, like modern medicine or ancient. It is not an exact science in many respects, though they do some wonderful things. Much of medicine today is not far above superstition, as we read in the papers, etc. An example was the Swine Flu tragedy, and many others. Read the book, Devils, Drugs and Doctors, by Howard W. Haggard, M.D. Many people tell me medicine is controlled by a large drug industry, and not by the practical needs of the people. They oppose any new ideas not doctor or drug oriented, just as Galileo and others were opposed, with Bible translators and readers burned at the stake. Remember in history how they used to burn persons at the stake thought to be witches, and I believe one was burned in the U.S.A. or perhaps just put to death. Perhaps the battle modern medicine has with the apricot pit and other natural schemes was brought on by themselves, by persisting in unworkable medical practice. The public will reject something after failure long enough. It appears cancer is chemistry, not a virus. New research seems to indicate cancer may be caused by a microorganism developed by chemicals in food and ecology. Not a normal chemistry.

If the vegetarian geologist is correct and organic plant life, over millions of years of time, made all the coal beds of the world why are boulders often found in coal beds? By what miracle or scheme, over millions of years, could these boulders be placed in a coal seam by vegetarian geology?

On page 100 of the October, 1977, issue of the Smithsonian Magazine, there is a very good article on the Amazon River and its ecosystem. If you could get a copy, read especially the material on page 110. Here is a short quote from this page: *"The time has come for us all to accept our responsibility as the implicit stewards of life on this beleaguered planet. Science will not save us, for by the time the living matrix of the world is understood, the brief interval allotted for solutions will have ended."* The article states further, *"This will apply to social, political and economic institutions."* The article says, *"The next 30 years should tell whether the great rivers and oceans will be lost to ecological destruction."* Earlier I quoted Dr. Paul Ehrlich, as he used a figure of thirty years for the life cycle and ecosystem including man himself. If, as science and evolution claim, it took millions, even billions of years to develop the earth's life and ecosystem, why is it possible to assign such a short time to bring about its end?

PRESSURE — DENSIFICATION

After several visits to the Petrified Forest of Arizona, the subject of densification, compaction, pressure, heat, and did these cause metamorphism, in wood fossilization or petrified wood? It seems they formed on or near the surface, and for the most part never did have a great overburden. A magazine published by the Museum of Northern Arizona, starts out by saying, *"Geologic time is divided into four units, called 'eras', each millions of years in length. That the Chinle Formation of Late Triassic age, is about 200 million years old."* This magazine associates the Petrified Forest with this age. Much of this area of the Petrified Forest was under up to 3,000 feet of rock and earth, it is claimed. Then it is claimed rivers and streams formed and carried away most of this formation. Then a cycle of erosion set in that stripped away the last of the Formation, cutting deep into the Chinle, revealing a large concentration of petrified wood and other fossils. My question is, can this be correct and did it take millions of years under pressure to form the petrified wood? Or, did they form right on the surface where they once grew before the flood of a few thousand years ago? Remember, petrified wood can be found in many places, and when I was a boy working on my father's farm, my brother and I found fenceposts someone had placed in sloughs where water stood much of the year, already mineralizing and heavy to carry, and impossible to drive a nail into.

Does it really take millions of years for wood to take on minerals?

It seems not. Page 6 of this magazine says, *"After 200 million years of burial, each year more of the petrified wood is exposed."* Further they say less than one quarter of an inch of soil is moved by erosion in this area now, each year, by water and wind. Most wind erosion though, will neutralize itself as it all must come down somewhere, and somewhere else is coming down here too. We must always contend with the law of displacement, even in water erosion. The above article says, *"Most of the eroded sediment now accumulates behind Hoover Dam, in Lake Mead."* Yes, that is true, and for this reason experts have fixed the silt life of Hoover Dam at 160 years. We could reasonably say dams like Garrison, in North Dakota, would have a shorter silt life, as here we have all earthen walls and drainage, whereas Hoover Dam has a large amount of rock structure. Fortunately, all laws physical and otherwise have built-in negation factors. Perhaps we could call this the Law of Retribution.

It is my claim that the Petrified Forest is not 200 million years old, and that it did not require all of this time for the wood to mineralize, nor was it ever buried as much as 3,000 feet underground. It would seem this great forest was growing just before the flood, and when the deluge came down, it destroyed the forest; and, for some reason, this area did not drain free of water for many years, soaking, evaporation, etc., and because the trees were under water for a long time, they simply mineralized, just as the fencepost on my father's farm did. Until someone comes up with a better explanation, one which is reasonable and provable, I will hold to this one.

May I call attention to a picture on page 24 of this magazine. There is a petrified log lying over an archway. Would you believe it has been there for 200 million years? Another place in the park there is a petrified log lying across a small ravine. It has since been supported by a cement structure to preserve it in that position for our viewing. Would anyone say this and the other, or for that matter, any of the wood had been buried thousands of feet underground and could still remain as it is? Did the tree fall over the ravine after it petrified, or did it fall there and petrify in the position lying across the ravine? Legend has it that cowboys once rode their horses across this log. It seems reasonable to say the logs petrified right where they are found lying, and nothing much has changed since they were first discovered September 28, 1851, by a U.S. Army Captain. Why does it always seem so easy to claim some thing or action took place 200 million years ago? Perhaps because no one can prove it.

It has always been a source of wonderment to me, in scientific circles, that it is always claimed that densification, solidification, hardness, molecular reduction, and all related theory, is only caused by great pressure over millions of years, plus heat, and a few other factors; then metamorphism sets in, and hence we have many kinds of matter and material and further the process never seems to end. I have never agreed with the idea. Is there a moratorium on reason and simplification? Is there any great difference between a diamond mined in a deep mine of South Africa or one found somewhere on or near the surface? Why is gold always gold if found as a nugget on the surface or a shallow mine, or gold mined 8,000 feet down, where perhaps pressure exists? Some petrified wood found on the surface is only two points softer than diamond! Why does one need pressure and millions of years to be what it is and the other does not? The wood mineralized right on the surface with no heat or pressure. I do have some petrified wood, which is very beautiful, on my desk. I bought it in the Park at a shop. I also have rocks from the Gold Mine at Lead, South Dakota, and dozens of oil well core samples from thousands of feet down in the Williston Basin.

For those who claim time, pressure and heat are the only factors to cause hardness or densification, please answer me this. I have been on dozens of oil wells drilling, mainly in North Dakota. At surface I have seen a 30 foot stand of drill pipe drill down in as little as 6 minutes. There are many stringers and strata of different hardness all of the way down. I have also seen 30 minutes spent drilling with 50,000 pounds of weight on the bit to drill one foot. The geolograph on a drilling rig records exactly how much time was spent to drill each foot to the bottom. If great pressure and other factors caused the earth to be so hard in one strata, why did it fail in its duty in other places and leave it soft, even very soft and granular? WHY? In some places we even find sand, and soft shales and sandstones, then we may find a very hard section, perhaps anhydrite. Oil drillers are always happy to find a hard section to drill through before they find the oil zone, as this means there is a good caprock to contain a fine oil reservoir. Ten minutes or more to drill one foot with from 40 to 60,000 pounds of weight on the drill bit would be very good for cap rock. Yet, if the drilling break into the oil producing area did not get much less, say three to five minutes per foot, we are very unhappy. We need a good porosity zone to produce oil from, much softer than that above the oil zone where oil and gas are trapped. My argument is, if heat, pressure, and time, all factors constant, made some sections in the earth so very hard, why didn't it do the same for all sections?

An example, if the vegetarian geologist claims that the crucible of heat, then pressure, and millions of years of time, caused things to be as they are, made all of oil, gas, coal, etc., by metamorphic action, would we not have a single environment at say the 8,000 foot depth in the earth? If, as they say, this is constantly operating, with constant factors, in a single environment, why shouldn't there be just a single element or structure with no variation thousands of feet thick? Yet this is not the case, and the absolute opposite is true. Great unexplainable variation in structure and mineral element exists.

If, as is true, there is always much softer and more porous structure under the hard layer above oil producing sand and strata, why didn't the same process claimed to have hardened the upper part also harden or densify the lower as it would be subject to the same pressure and heat and other factors they claim produced the solid, tight upper section.

Perhaps we have been influenced by standardization of thought and theory too long, from a wrong premise. It would seem to me that most of the geology of the earth has been just as it is for thousands of years, and perhaps most of it is today just as it was laid down, with little or no change, other than the surface arrangement caused by the Deluge. Diamond was always diamond, metals were always metals, oil and gas were always as they are found, as coal, etc., is just as it was put into place. Fossilization in coal and rock proves the materials has been undisturbed since it was put into place, with little or no metamorphism. The argument is not how could this be, but rather, how could it be otherwise? True, the crucible of heat was used by the original igneous earth in the annular systems, in millions or even billions of years ago, as some say. We know what heat can do to metal to make it hard, like the impeller shaft and fins in a jet engine. So over all of those years, when the creative process was active, we did not have a single environment continually. As material was cast into space and later returned, each in its own time, this accounts for the layer upon layer of earth's substance and would allow for great variation in elements.

Why do scientists ignore the law of displacement? There is only so much of anything. If it is not here then it is over there. If, as they say, it traded places, it has to be accounted for. On the one hand, it is claimed that something was buried under thousands of feet of rock and earth. On the other hand, others will claim something was uncovered from thousands of feet of material. Where did the energy and force come from to do these great tasks? Did the same force uncover this area as the one that covered it up in the first place? Where was the material they claim covered it at before it did its job? Then, when it was uncovered, where did the material go to after it was removed? No action is automatic, all actions require a force which is energy. Gravity is an unending force, yet once it performs its duty and a job is finished, it is now neutral, and this force no longer will move the object, until an outside force is applied. It would seem that things in the last several thousand years have not changed all that much as some claim, and perhaps most things in the earth are just as they were when they were put into place originally.

CHAPTER 4

SPONTANEOUS COMBUSTION

We cannot violate the Law of Principal, and take up new positions until we account for the old ones, with an orderly chain of facts, without any missing links. Let us consider some thought, mainly from Vail and Dana (U.S. Geologist & Mineralogist, 1813-1895), though others do contribute to the philosophic picture. Let us build on one subject, spontaneous combustion. Sometimes there is a temptation to move from position to position without proper connectors. We may assume we will not allow ourselves to do this.

Spontaneous combustion provides a key as to how carbons came about. For example, much coal is pyrophoric, it can ignite spontaneously upon contact with oxygen. Russia has much of this kind of coal. Now, if the vegetarian theory of creating coal and carbon, or hydrocarbon, in an environment of free oxygen necessary for plant, animal and later humans to live in, how did it escape combustion when it had ample opportunity? Would it not self-destruct in its own process? When plants and animals die, are they sealed away, to escape the devourer (oxygen), and become an unspent fuel? Either the igneous earth formed all of the carbons, or vegetation and animal did. There is no in-between ground to stand on. If vegetation formed carbons before, then why are they not doing so today? We must conclude, carbons came into being millions of years before plant cells did. The igneous earth did not need direct sunlight nor the mysterious laboratory of plant life to lay down the carbon beds of the world. The primeval, primitive carbon beds in this process would also include graphite as well as peat. Geologists refer to peat as new coal, meaning it has not had time enough to metamorphose into coal. It seems when geologists and scientists have no other explanation for some action, they use the overworked word "metamorphism".

Are we to conclude the elements of the earth are only a combination of other elements by metamorphism? Why is it if a fossil, such as part of a tree, happened to extend through a coal seam into the clay above, it is mineralized as carbon in the coal, and mineralized with clay in the clay? If vegetation made the coal, why did it not become coal in both places? As a matter of fact, it is not coal or clay in either, but a foreign material in both.

Let us argue as Vail did! Let us admit that by some unseen and fortuitous means the original carbon fuel was afterwards metamorphised into plant food, and was eventually transformed into fuel. This will lead to a dilemma, so let's proceed anyway. If coal beds are mineralized vegetation this would necessitate an ocean of oxygen. Dana, seeing this inevitable fact, said, *"The atmosphere now contains less carbonic acid than it did at the beginning of the carboniferous period by the amount stored away in the coal of the globe,"* and yet the same high authority says, *"much more carbonic acid (than now exist in the air) would be injurious to animal life."*

We know some coal beds are a few feet thick to 300 feet thick. Some experts say if all of the coal of the earth were moved to the surface, we would have a hypothetical layer 10 feet thick around the globe. Be that as it may, we know plant life cannot make carbon out of nothing, so if we remove the carbon from the earth, and leave it to vegetation to create it, where was its source of carbon to build from? As Dana suggests, it had to be transferred to the atmosphere to fulfill the process of vegetarian geology. Here is our dilemma: Plant and animal must live in an atmosphere of free oxygen, though plants need less oxygen, as they use carbon dioxide and give off oxygen, yet in your wildest imagination, can you conceive in your mind what kind of atmosphere we would have if all of the carbon in the coal beds of the world were transferred to the atmosphere, to be used by plant life to transfer back to the earth, all of the coal beds we now know exist there. Would you say animal or man could breathe such an atmosphere? Perhaps it may be so thick even mobility would not be possible, nor could we even see our hand right in front of our face. It is absolutely evident no man or animal could live in such an atmosphere nor could plants then either. As coal is sometimes very deep in the earth, men of science say this process went on for millions of years. Can you believe this is possible? In a conversion estimate one expert suggests the coal of the earth would create a carbonic acid layer over the earth into the thousands of feet up, redensifying the atmosphere by billions if not trillions of tons, far in excess of our average 30 psi. What form of life could live in such a mythical world?

It is true, plant life did precede animal and then man in the creation process on earth, and plants did help clear up the atmosphere and bring about a good oxygen-carbon dioxide balance for animal and man to live in; however, there just is no way plants could have activated the impossible vegetarian process herein described.

Could even the bacteria world live in such an atmosphere we here set up? Bacteria life is essential in recycling the elements of life, and its main duty with higher forms of life take place after death, in the process of recycling elements, not to coal, but back to the same things again. Therefore, there is not scientific evidence of any metamorphic action, to any great degree, in the elements of the earth in recent geological times. Too, spontaneous combustion disproves vegetarian geology. It is logical that carbons of the world are simply unconsumed fuels left over from the igneous earth, and were created long before life was. This fact was brought to my attention years ago when our home burned down from a chimney fire caused by soot collecting there as an unspent fuel, just as the old igneous world had done millions of years ago. As one expert put it, the carbon comprising the peat bed is simply unconsumed carbon.

My challenge to the chemist and the scientist is this: That there is no evidence of any metamorphism in any of the basic elements of the earth in recent geologic times. (I do qualify this by basic .) I am not including actions by the catalyst, the atom smasher, or genetic manipulations and gene splicing, nor do I refer to solid state fuel in rocket engines, nor normal, natural, radioactive decay. Recombinations are not new elements and life forms, but only an unnatural recombination of existing forms of things. Had original creative process built in many metamorphic possibilities, all creation would some day dissolve into a single universal element, and too, without the absolute genetic block, all life forms would reduce to one.

Isn't our world described above nothing more than a system of pollution, the number one problem in our day, threatening life itself? Will such a negative process evolve into a clean, paradise world as we now have lived in, until man-made pollution began to take over? As to the bacteria world we cannot live without, isn't it, too, being destroyed slowly by chemicals, including fertilizers which destroy the life sustaining bacteria in soil. I am told most chemical fertilizers are hormonal in action and not plant food. The process may be compared to having a very nice bank account, then writing checks on it faster than we replace the funds until bankruptcy. If we slow down on writing checks it will last longer, though the same end result. At first, fertilizer hormone action does seem to produce more growth; however, we are spending our soil fertility faster than it is replaced, until finally bankruptcy. My point is, if the ecosystem we now live in is so delicate, it may be so easily destroyed, how could the carbonic acid environment Dana mentions bring about such prolific plant life and animal life to make all of the carbons of the earth? In this vegetarian theory, as considered, if only hypothetically, would we not need to multiply the negative factor by the many thousands? Perhaps we have made our point.

ENERGY SOURCE

A Particle Physicist on the Donahue Show, April 10, 1980, said, *"The world exists of particles and the forces between them."* His explanation of the material universe was excellent. Yet, it seems the deeper they go into smallness, the more unanswered questions come up. Likewise, the more we find out about outer space, the same thing happens. No one really knows, do they? It is true the only constancy we have is everything is constantly moving, as atoms are in mass material. Atomic science was born with the ancient Greeks and possibly the Romans, who would say and argue, if we had a small piece of gold, and a tool with which to keep dividing it, could we keep up the process until finally there was nothing left.

Then they would argue it is unthinkable as we cannot make something into nothing. Conversely, we cannot make nothing into something. With modern technology, all we can seem to find out is that all material substance is, is a series of electrical charges, negative and positive, with neutrons. Particle physics is another subject too complicated to get into now; however, I will quote one article on it. Attempts to understand what our universe is made of has produced the science of particle physics. There is a search on for the ultimate particle. Some conclusions were summarized in, "Science Digest Special Quote": *"If there is any ultimate stuff of the universe, it is pure energy,"* further, *"but subatomic particles are not made of energy, they ARE energy." . . . According to particle physics, the world is fundamental dancing energy. "Governing this energy, and limiting the forms it can take are a set of conservation laws."* It will not change what we have discussed so far, anyway. The answer the Greeks did not have, is that energy and mass are exchangeable, convertible, transformable in relation to each other, things are one or the other.

An interesting observation is when a young college graduate, a physicist by the name of Albert Einstein — who could not find a job — offered his new theory. He expressed his belief that there was a relationship between matter and energy; frozen energy. In Europe, before a university, he described his simple equation as, "E" is energy, "m" is mass weight, "c" he calls the speed of light "2" squared. He explained that the speed of light is 186,000 miles per second, now if a small amount of matter could be converted to "frozen energy," the energy release would be enormous, perhaps an ounce would be equivalent to the explosion of a million tons of TNT. His mathematics impressed his colleagues, though they felt the possibilities were preposterous. The idea was compared to the medieval alchemist who dreamed of changing lead to gold. As is often the case with something new, many felt the theory would be disproved. Ein-

(47)

stein knew the equation was correct, and he wondered if the energy could ever be released.

Nicolaus Copernicus shocked the scientific and religious world when he advanced the theory that the sun is the center of the planetary system, which became the foundation of modern astronomy. At first, the idea was rejected. Next, a man, Niels Bohr explained the atom as a small duplicate of the solar system; a nucleus (sun) surrounded by electrons, as orbiting planets. Gravity holds the solar system together just as negative and positive electrical charges hold the atom. Soon the secret of nuclear chain reaction was discovered. The famous Manhattan Project was set up at Los Alamos, with J. Robert Oppenheimer as laboratory head. The first atomic bomb test at Alamagordo, New Mexico, was a success, depending on what is called a success. Now there is the Hydrogen (Thermonuclear) Bomb. Most use fission-fusion-fission devices and are many times more powerful than Atom (Nuclear) Bombs.

Oppenheimer, as well as others who worked on the Manhattan Project for the first atomic bomb test at Alamagordo, was stunned by the sheer magnitude of the blast; according to information in the above two paragraphs with some thoughts from "Nuclear War — What's in it For You?" (Why Do You Feel Scared With 10,000 Nuclear Weapons Protecting You? From the cover of this same book.)

The above mentioned book says, *"This story did not have a happy ending."* I once read an account of this test at Alamagordo where some said a few people involved in the test thought the chain reaction may not terminate as planned and blow up the earth. You know the rest of the story.

We will be hearing more about the ghost elemental particle called a Neutrino. It was once thought that neutrinos did not have any mass or weight. If neutrinos do turn out to carry mass, some feel they will become the dominant material of the universe and resolves some long standing mysteries. One physicist said, *"It's incredible, but not inconceivable."* Answers may pose more questions, as usually is the case.

Energy must come from some source for every action, as an earthquake, needs energy or force. Where did it come from? It is gravity, a constant force. The deluge forced the ocean floors down some, and heaved the mountains up some, this caused the faults in the earth, which in turn are only settling toward a neutral position because of gravity. When they all find their neutral place, there will not be any more earthquakes. Not too hard to understand is it, as more answers are found in simplicity than in complexity. From this point of view, it seems that 90% of everything we have been taught, and feel is correct and true, is not correct or true and oftentimes the absolute opposite is true. Many who do not think for themselves will not like this, or agree. I am not one of those. Let us just take a very few examples. What about the Piltdown Man as taught to be a missing link in evolution. That has been proven to be a complete fraud. Another is, we were told electricity generated from nuclear plants would be so cheap we would not need meters. Now they say it will be the most costly kind of energy. Because of cost overruns on nuclear plants, a '60 Minutes' program pointed out this would be very high cost energy.

So far, fission reactors (nuclear) have been beset with woes — financial — and other unsolvable problems. Therefore, it will be difficult to sell a fusion program to the utility companies. Fusion has been a costly program for many years and they have not made even a break-even of energy used to energy consumed. Many have grown wary of fusion — who knows if there is some yet unknown instability affecting magnetic-confinement nuclear-fusion, ending the entire scheme in fiasco? Radioactive danger may, in fusion, turn out to be greater than with fission. Time and money will decide.

MUTAGENS

What about mutagens, mutations, mutants, or to artificially mutate? All forms of microwave radiation are mutagens, artificial not natural. This could destroy life on earth in time, and has caused great life reduction already. To artificially mutate genes does not result in a stronger life form, but rather a weaker life form. Yet, these are the building blocks of evolution. We have been taught that life forms are upgrading constantly, when again the opposite is true. Chemistry has not been a blessing in this area, as chemicals are dangerous and causing no end of problems, with even the waste they leave causing death and disease. Chemicals, too, can cause mutations, as we saw in the use of the tranquilizer drug thalidomide. Where do we see any upgrading of life here or a new and better form of life? Siamese twins are mutants, as are many other unnatural deforms. In the book, the New You and Heredity, by A. Scheinfeld, he adds, *"It is through the rare instances of favorable mutations, of innumerable kinds and in countless numbers, occurring successively over very extended periods, that the whole process of evolution may now be explained."* Are we to conclude that a series of accidents and mistakes in nature, perhaps caused by artificial means, is how the marvelous life forms on earth came about? This just cannot be the case. The opposite is true. One very early problem caused by this goes back to Roman times. Rome used much lead for water pipes and many people, particularly the poor, used lead drinking cups, etc. It was a great factor in the destruction of ancient Rome, even though they did not know the cause was lead poisoning. Yet, it was a simple thing, lead was easy to work with, and cheaper than other metals.

It was the lifelong desire of Albert Einstein to fit all of the forces of nature into a unified set of equations. He held firm to his faith that order, not chaos, ruled the cosmos — for, he said, *"God does not play dice with the universe."* In considering the Scheinfeld explanation of evolution here, it would seem his entire premise is nothing more than a game of chance. He is playing dice with the creation. Not only that, his odds are in the billions if not trillions to one. His theory based on chaos, is incomprehensible. In the widely accepted teaching of modern geology and other scientific philosophy, isn't it just a case of the triumph of propaganda over truth? There is a saying in legal circles, *"The burden of proof rests with the accuser."* Anglo-Saxon laws says, *"We are innocent until proven guilty."* Roman law says, *"We are guilty until proven innocent."* In the case of the claim made by science that oil and coal are the remains of fossils which bacteria lost to a landslide or swamp, why not let the burden of proof rest on the claimant. In the court of reason, it is now time for the final argument. Where is their proof?!

WORKABLE IDEAS STILL IN USE

Why must we always look to some deep complicated answer, when most great things and accomplishments start from a simple and humble beginning, with little or no change except improve on the original idea. Two inventions helped make the oil industry what it is today. First, the father of Howard Hughes invented the rotary drilling bit, making it possible to drill miles into the earth through a surface pipe cemented in the earth, instead of working with a cable tool, with open hole, and a bailing system, which would blow out when high pressure was encountered. Second was the catalytic cracker for refining, a great improvement over the thermo cracker or the teakettle, first used. Thousands of products may now be produced by the catalyst method. This, though, is where the chemical industry problem began. Man did not stop soon enough.

Great industry is built on a simple idea with hard work, and the idea is still in use by most who want to compete. Look at Henry Ford's Model T Car. The planetary transmission used in the Model T Ford is still in the modern automobile, the only difference is, it is now automatic. The planetary reduction gear Mr. Ford had on the steering wheel of his Model T is just like the one used in all four wheels of a four wheel drive tractor today. Have you ever heard anyone give Henry Ford the credit for these great inventions? There are many more such cases no one ever hears about, perhaps someone else even stealing the idea and the credit. The entire universe is made up of both simplicity and complexity, simplicity should and does predominate. Henry Ford, as with others I am sure, had no idea what his company would grow into at the time he started it. So it is with many things.

Another example of wrong information, many of us were taught in school, is Columbus discovered America and proved the earth was round. Reader's Digest, November, 1966, has an article showing the Vikings discovered America, even though people already lived here at the time. Alexander the Great, a world ruler who had many scientific men working, made many things and ideas we still use today. In the third century B.C. they knew the earth was round. At that time, Eratosthenes, using accurate distance, gave the figure of 26,000 miles. The actual girth around the earth is 24,901. That was very close. Of even greater interest are the words of the Prophet Isaiah, who finished writing the Bible Book of Isaiah in the year 732, B.C.E., where he tells us the earth is a circle. Quote: *"There is One who is dwelling above the circle of the earth, the dwellers in which are as grasshoppers."* Isaiah 40:22.

Ptolemy, the great astronomer, put the earth as the center of the cosmos. He reasoned if the earth moved, objects cast into the air would be left behind. He was wrong. The earlier Greeks knew better. Archimedes fathered the endless screw (now called an auger), to lift water to a higher level to water land. It is still in use. His perceiving the enormous potential force of the lever and fulcrum, he boasted, *"Give me a place to stand, and I will move the world."* Ptolemy's theory the earth was the center of the cosmos held sway for 15 centuries. Galileo (1564-1642), a great thinker and scientist, said the Milky Way was a mass of stars too numerous to count, or beyond belief. He also said there were sun spots, and supported the idea of Copernicus, that the earth moves around the sun. Aristotle did not agree, nor did the Church, and such books were prohibited for 200 years by the Church. For this he was summoned before the Holy Office, or Inquisition, and forced to give up his belief in the Copernican Theory. He was sentenced to an indefinite prison term, and finally banished to his villa.

The great bodies of religion then, as well as now, are not much more than superstition. False theory and ideas cannot be kept alive for centuries anymore, as computer research has replaced blind-alley research and superstition. Just this little information would make it necessary to rewrite many history books. History will show man's origin was Mesopotamia. How did he get to this part of the world, America, before Columbus or the Vikings? National Geographic has had some good articles showing they could have crossed the oceans in a reed raft or boat, to Central or South America. Are some of the things men claim all so monumental? Perhaps all we need is some simple truthful answers.

EVOLUTION

Perhaps I should comment on one more subject further than I have. There are many experts to quote from, but I will base my remarks on a book written by very prominent Cornell University professor Dr. Carl Sagan, author and writer of many books and papers. The book is a best seller entitled, "The Dragons of Eden." If he proves one thing in his book, it is that the universe and all of the forms of life on earth are so wonderful and unexplainable, they could not be an accident, or just evolved from nothing, but had to have a great Master Mind behind all of their perfection and beauty. This man tells how wonderful the creation is, in this he is correct and can prove it. It would seem to me that God is the greatest scientist of all time, in that He created all things, including life itself. Darwin, Wallace, and Sagan would disagree. That is their privilege. On page six this remark is made, *"Evolution is a fact amply demonstrated by the fossil record and by contemporary molecular biology. Natural selection is a successful theory devised to explain the fact of evolution."* This is absolutely not true or correct, and there is not one shred of evidence to prove them. It seems when someone is famous and has many Doctor of Degrees, etc., and studied in great circles, there is a tendency to believe what they say without proof. I have years of experience in geology and oil drilling and find no such record in the fossils.

On page 13, he is on the subject of cosmic chronology, or cosmic calendar. Why do many such men build in so many disclaimers, perhaps, maybe, if all these things can be true, thus and thus must be. *"But we have been preceded by an awesome vista of time, extending for prodigious periods into the past, about which we know little — both because there are no written records and because we have real difficulty in grasping the immensity of the intervals involved. Yet we are able to date events in the remote past."* As support he used geological stratification, radioactive dating, archaeological, paleontological, and geological, as well as astrophysical, note, as theory only he says. Then he says, *"The Chronology corresponds to the best evidence now available. But some of it is rather shaky."* We did not find one absolute in all of this, did we? One may as well say, your guess is as good as mine. Not even one of it was provable. The author says knowledge of our destiny is our destiny. TRUE, but where will the knowledge come from, and who will decide? One chapter is on the Future Evolution of the Brain. Yet the truth, which we can all prove, is man has been degenerating backwards from his beginning, physically, morally, mentally, and in intelligence. This is opposite to the Darwin theory of natural selection he supports. When the famous and intelligent Dr. Paul Ehrlich, and also an author, from Stanford University was asked on the Tonight Show how long he thought human life could last on earth at the present rate of pollution and ecological destruction, his answer was 30 years. We would agree with Dr. Ehrlich as we are aware of what is going on now all over the earth. The question is, when will the turn around be?

This book has many pictures of half-men, half-animal creatures in it. We know someone had to draw them from imagination as no one has ever seen such a man. Nor has any fossil ever been found to show one. Many other authors and books of science show the same kind of pictures. Time-Life put out a lot of books on the origin of life with such ridiculous creatures in it. How any sane mind could believe such stuff is beyond the power of reason. They must be the figment of hallucinations, pure lunacy. It may be compared to Dante's Inferno, no doubt dreamed up under hallucinatory spell. No such creatures ever existed. They are nothing more than imagination by the artist. A small piece of bone fossil may be found, and they build this entire man-animal creature around this small bone. Not only that, they feel they can account for millions of years of human evolution by this one little fossil and just fill in the rest as they see fit. They even try to say they were right-handed or left-handed. Who can believe such stuff?

There is a very interesting article in the June, 1980, issue of the Smithsonian Magazine by John Pfeiffer, entitled "Current Research Casts New Light on Human Origins." I do not criticize Smithsonian for publishing this article, as they are just publishing the ideas of other people. In fact, I am happy they did publish this material as it gives us a lot of subject material wherein we and I are able to make good logical decisions. We must never refuse to examine all aspects of a subject. There are pictures and charts, none of which can be accurate, only someone's guess at best. Many authorities on this subject are quoted including Darwin and the Leakey family. Many long technical sounding names are given to various ape and similar creatures, perhaps for identification, and mostly to impress the reader. One writer goes on and on explaining, in detail, how things were 15, 10, 5 and finally 3 million years ago just as though he had been there walking around and made a firsthand examination of the situation, of these prehuman men as they call them. The detail is so exact and precise, it is laughable.

The chart show the line of descent from Australopitheous to homoerectus and then to homo sapiens (man). The scale starts at 40 million years ago and ends at zero, or man. None of this is provable nor does it make any sense to me. This article brings in some fiction (which it is anyway) by a French novelist using the pseudonym Vercors, published a tale "You Shall Know Them" and the "Murder of the Missing Link", which seem to make as much sense as the main body. One statement says a certain creature is no longer among us in a recognizable form, as he eventually evolved into human beings. Can you believe an unprovable remark like that? Then they go into an un-proven "protein clock" by injecting human albumin into chimpanzees. They feel this will prove every creature evolves and still does. Then they go into the matter of molecular biology, claiming biochemical evidence will help them prove a point. They say evidence from molecular biology as well as morphology is proving their theory. Also, it is claimed, deep DNA research will be the key to unlock evolution's mystery.

Up till now, DNA research has not simplified their position, but only compounded the problem. One item of interest to me is so many anthropologists and scientists are saying the transition from ape to man took place about 500,000 years ago, rather suddenly, that is in comparison to the 40 million it took to get to this point. So many making the claim of 500,000 years ago, doesn't this violate due process? Where in molecular biology do we see sudden change? Isn't this against their own theory? Why should there be a short span transition in a process they claim only can sustain itself through eons of gradual time? It is folly to try to make head or tail out of such unprovable, farfetched claims. May I ask, has anyone ever proven a specie change, kind to kind? Words are cheap, but is there any factual proof? No there is not!

I do not have a quarrel with evolutionists. It is not my wish to get them to abandon their ideas and theories. Who cares? All we ask is let them prove their positions and claims step-by-step, without contradiction and inconsistency. Or, worse yet, no proof at all. Too often their own concepts undermine their own stand and credibility. If they do not come up with factual, provable, truthful information, we are forced to look elsewhere for answers. The entire spectrum of creative process from the scientific and evolutionary process are inconsistent, and contradictory. Their own documentation is nothing more than fiction. Just not provable. Perhaps a line in the June, 1980, Smithsonian will give us a clue as to why a futility is pursued so ardently, it says, *"The find made headlines, and ample financial support soon followed the publicity."* If you can get a copy of this June, 1980, Smithsonian Magazine you will enjoy reading the article I refer to. Then you will know what I mean about unprovable claims.

Regardless of the claim of paleontologists, anthropologists, geologists, archaeologists, the entire spectrum of science, the origin of man on earth cannot be pushed back into the millions of years. They do not present proof. No matter how hard the blow is to them, their testimony will not stand the pressure of facts. Conjecture and rhetoric are used for facts. Taking every possible evidence we have into account, plant life appeared on earth first, then the animal kingdom, some of which served a purpose and passed off the scene, then man was last to appear. Fossil record will support this, as well as true science. Vegetarian geology is not a true concept. Let us remember the same kind of men believed the earth was the center of the solar system for centuries. So it is with old accepted dogma. Rather than ape make a sudden transition, there must be a mental transition.

I have just gone through Scientific America on Human Ancestors. The material is all about the same as the Smithsonian I refer to. On Homo Erectus, one author says the transition to man took place about 500,000 years ago. Much is said about evolution of human walking. Also the evolution of the hand is covered. One writer says man's brain began to differ from that of other primates some three million years ago. There are skull pictures, etc., yet all of this does not prove anything except all things are about as they were, nothing changing, maybe a little in size and that is about it. In going through a number of museums of natural history, I would say in skull displays, even an amateur would see the absolute line of demarcation from ape skull to man skull. Where then is the evidence of transition? Man has always had feet, arms and legs and could walk upright, as do animals not related to man have needed body parts to live in their world. Plant life had to precede animal life, and then man, to use up a lot of excess carbon dioxide present, and release oxygen, for man and animal to breathe.

Through photosynthesis, carbon dioxide and water are absorbed by plants which synthesize other elements and release oxygen into the air. Once a normal balance was established, man, animal and plant can live together forever. However, man is now destroying the oxygen balance with pollution, using oxygen faster than the process can create it. Plants, man and animal have not been on earth many millions of years as claimed. The oxygen-carbon dioxide process is just not that old. Life had to follow it, not precede. Oil had to precede all of this, as much free oxygen as we now have will not make organic matter into heavy carbon, but rather decompose it for recycling in the life sustaining process.

On page 53, Dr. Sagan tells of a small fish swimming in a primeval sea 500 million years ago. He claims there is a little swelling at the front end of its spinal cord, this is its brain he says. Therefore, he claims we came from that little fish 500 million years ago. Can you believe such a thing? Men like L. S. B. Leakey are quoted and Richard Leakey, and many others. Animals have always been animals and men have always been men. The Leakey family cannot prove any of their wild claims as to the origin of man, nor the age of man into the millions of years. The material is just invented.

Take the Early Relatives of Man, by Elwyn L. Simons: *60 to 12 million years ago, evolution singles out the ape stock from which humans arose.* The same writer says, *"Ramapithecus has clarified its place in human evolution."* Next we have a subject, The Evolution of the Hand, by John Napier, Tools found more than a million years old. Same author, on Antiquity of Human Walking. Homo Erectus, by William W. Howells; he says it seems possible the transition from ape to man took place some 500,000 years ago. Next we have Stone Tools and Human Behavior, by Sally & Lewis Binford. Then another author spells out, The Functions of Paleolithic Flint Tools. Still another writer writes on The Food Sharing Behavior of Protohuman Hominids. The list is endless. Some of this material is interesting to read. All they are is excellent fiction writers. Absolutely none of their claims are provable beyond a few thousand years ago, yet one has gone as far as 60 million years back. May I ask these scientists who make such claims, how they prove which factor was in force at any given time, erosion or sedimentation? One group claims erosion is a constant factor. Others claim stratification by sedimentation is a constant factor. Where do we draw a line? If we say one is true, then these items and fossils should be buried miles underground. If erosion held sway that long, there would be nothing left of the evidence. Which leg are they going to stand on, or perhaps they do not have a leg to stand on. Others will set up a tug-of-war by claiming both of the above factors are constant at all times, so they can change back and forth in an argument and use both at the same time. In my lifetime, I have learned one great truth, the weakest and most ridiculous position anyone can take, is that of inconsistency. This is a powerful test tool.

TRANSLATIONS

I have drifted into many other subjects and have not kept them all in order as I should, though it seems necessary to make many comparisons to prove a point. When one starts on a subject so interesting, such a flood of thought and ideas come to mind it would be easy to write a book on each subject. It is a wrong impression that only those with a supermind can understand, or figure out, these subjects. This is misleading and years of education are not needed to learn facts. Always remember, fact and truth stand alone, and do need opinion to validate them. The number of people who believe or do not believe does not add to, or detract from, fact. It stands immovable regardless of what public opinion is for or against. Given time, the compensatory law of retribution will grind any who disagree with fact and truth into the ground. A simple example in translation of the Bible by many English translators is the one single English word "world", it is used to translate four different Hebrew and Green words. So when Christ said the world would come to an end, he did not mean the ground we walk on. Sometimes we are led to believe that languages and translation are too deep to understand, and the books brought down to us by translation are not correct, the Bible is especially singled out for this criticism. This is not a Bible thesis though I am going to mention some points to prove a point. That is their privilege and regardless of what anyone may believe or not believe, it will not change anything from what it actually is. We have the physical evidence before us at all times. Tens of thousands of Bible manuscripts and parts of them exist in many languages, including stone and on many materials for writing.

Yet, all of this is not all that complicated as we may be lead to believe. Moses began compiling the Hebrew Text in 1513 B.C.E. (Old Testament). Because many Jews were brought to Alexandria, Egypt, there was a need for a text in Greek, so the Greek Septuagint (about 280 B.C.) was made, the first translation of the Hebrew Scriptures into the Greek. Now there were both Greek and Hebrew translations of the Old Testament as many call it. Matthew finished the Bible Book of Matthew in 41 A.D., the first of the New Testament as some call it, or Greek text. He wrote it in Palestine, the popular Hebrew of the day and later he translated to Greek, as was all of the rest of the New Testament, or Greek Text. Many vellum manuscripts are still with us, both Hebrew and Greek. Then there are many, many valuable manuscripts such as the Vatican Manuscript, 1209, Vatican Library. Also the Sinaitic Manuscript, fourth century C.E., British Museum, London. Also the Alexandrian Manuscript, the Greek containing most of the Bible, also at the British Museum. The Dead Sea Scrolls have been great help. This brings us down to our day, with many popular translations, including the King James and Catholic Douay translations.

We have not scratched the surface on this subject, yet with this basic information, it is not all that complicated nor do we have many languages to go through. Translators today may go direct from Hebrew to English, and Greek to English. No problem at all. Keep in mind the archaic aristocratic style of words used in the King James, Douay, and many other translations of the Bible, are not any more sacred words than the modern English words we now use. They are the kind of English words used in England in the 16th century. At that time they understood these words, and today we do not. As in I Corinthians 10:25, a meat market is called a shambles. Someone was heard to say, foolishly, of course; if the King James Bible was good enough for the Lord, it was good enough for them.

At this time, we could do well by reading one of the great speeches of all time given by the Apostle Paul, before the court of the Areopagus, in Athens, at Acts the 17th chapter. Then we have many history books, one of the best by the great historian, a Jewish man, Flavius Josephus. Really, we are not lost in a no-man's land without information. It is everywhere for the taking. No one, great or small, has a corner on it, not even the intellectuals. In fact, they may get carried away with their own interpretation on matters instead of allowing the facts to answer, and stand, as they are. As has been said, facts do not need the approval of opinion.

Sometimes the answer to a puzzling situation may be in the element of time, not allowing enough of it. If we conclude too soon, before enough facts are available, accepting some authority in haste because it is popular, we may be wrong. An example, Bible critics once questioned the authenticity and historicity of the Bible because a long list of Kings mentioned in the Bible were not accounted for in secular history. They were grouped as folklore and fables of the Old Testament. A book "Dead Men Tell Tales," by Harry Rimmer, shows they have been all accounted for, through archaeological diggings and other methods. This book mentions 47 accounted for and I have heard of even more. Is it not true that science and medicine, 20 years later, found they were killing, not curing people? Time tells the truth. Another subject comes to mind I should have put in somewhere else in this thesis though, the popular nowadays idea of gene splicing, called genetic engineering. It is a technological trend, some putting a patent on them. A redesigning of life forms by genetic manipulation, an alteration of life form by genetic recombination. This was the subject of the Donahue Show, May 14, 1980. I will predict this will not turn out to be a blessing, but rather a curse, given time. As we said earlier, most accidental mutations did not result in an upgrading of life forms. This could be compared to the use of catalyst to create compounds not found in nature, and far more dangerous to life. As always, time will tell. Then, as usual, it will be too late.

Modern archaeological work has uncovered amazing amounts of information valuable to history, helping us to know man has not lived on earth near so long as paleontologists claim. We are getting a more accurate understanding from this work. Until the excavating work of Yigael Yadin in about the 1960's, the only history account we had of the great fortress citadel, Masada, built by King Herod, The Great, was that of brilliant historian Flavius Josephus. Josephus gave a very accurate account as the excavations reveal. In 73 A.D., 960 Jewish Zealots took their own lives rather than submit to Roman capture, when they felt they could no longer defend the citadel. The work of Yigael Yadin, MASADA, Random House, with the pictures, makes very interesting reading. Also, "The Bible And Archaeology," by J. A. Thompson, Erdmans is very interesting. Other interesting publications are American Schools of Oriental Research, Biblical Archaeologist, Smithsonian Institute, National Geograhic, and many more.

Yet, there are too many language barriers, too many provincial boundaries, too many religions, too many political divisions, all causing hatred among men. Abraham Lincoln once said you cannot help the poor by destroying the rich, nor help the weak by destroying the strong. This seems to be the case now. As stated before, Darwin's theory of natural selection is not true and the opposite is taking place. As Dr. Paul Ehrlich said on the Tonight Show, everyone feels their little bit of poison or pollution will not hurt, yet when this is multiplied, and in combination, it can be lethal. He also said, all many of the huge man-made projects, including some dams, are, is a WPA project or government subsidy to high paid engineers and scientists. May I add that this would not be so bad, if it were not for the destruction and misery they leave behind, for the next generation and sometimes the same one. William F. Buckley, Jr. said, on the Donahue Show, to a question: *"There is a chess game, the first move has been made, now you take over."*

CHAPTER 5
GEOLOGICAL STRATA & NAMES

Back to geology again. No one knows all of the answers now and perhaps a thousand years from now much more will be known. There are a number of main geological ages by name, a few are Mission Canyon, Madison, Red River, Winnipeg, Devonian, Mississippian, Ordovician, Cambrian, Pre-Cambrian. Geologists have found many identifiable sections within many of the main geological structures and given these names as well. One is the Deadwood sand (Cambrian) after Deadwood, South Dakota, where it outcrops. Cambro-Ordovician, Paleozoic rocks in the Keene area. Winnipegosis, Niscu, Interlake, part of Silurian. This goes on and on. The term Nesson anticline covers the entire spectrum of ages in the one area of the Williston Basin of North Dakota. A good thing we do have geologists.

It is not my purpose to be critical of geologists. It would be next to impossible to drill a technical oil well today without at least one to watch over each well. They are very well trained and experienced and intelligent. I have known and worked with dozens of them on oil wells. Like all imperfect humans as we are, they too make mistakes once in awhile, this may be expected. We all do. The Williston Basin is a very complex geological area, the geologists are doing a superb job of developing this three state area including some of Canada. To show how complex it is to find oil, it is sometimes like looking for a needle in a haystack. The Clarence Iverson discovery well, in April, 1951, was completed as the first producing oil well in the state of North Dakota. The Iverson was completed as a Silurian producer at 11,660 feet deep. Later it was decided to go back in the Iverson well and set a retrievable bridgeplug below the Madison, set a perforation gun, and open the Madison for production, which is just below 8,000 feet here. It made an oil well, so it turned out they had drilled through an entire oil field and missed it. This could easily happen though.

This will be supplemental information on the Clarence Iverson discovery well in North Dakota. The North Dakota Geological Survey, and Amerada Petroleum Corporation records helped me. The Iverson well was completed as a discovery in the Silurian zone 11,630-11,660. On April 5th, 1951, the well was acidized with 4,000 gallons of acid and it made a well. Later the well was plugged back to the Devonian-Duperow, 10,490-10,530 feet. This zone also made oil. The Bakken well was drilled north of the Iverson a few miles and was the first Madison producer. Following this completion, the Clarence Iverson was plugged back to the Madison and made an oil well there. As it turned out, the Iverson well went into three producing zones, and discoveries were made backwards.

Did you know that after 28 years of production, the first well to produce in North Dakota, the Clarence Iverson, Williams County, has been abandoned. The well produced an accumulative from three formations of 536,489 barrels of oil. A new well may be drilled.

They have been drilling Madison wells ever since and, perhaps, will be for another 50 years. My disagreement with vegetarian geology has nothing to do with them finding oil, but how it got there in the first place. Oil is where you find it as all geologists will agree and all that will find oil is a drilling bit and enough money to back it up. Keep in mind the above mentioned wells required a large amount of acid to make a well in the tight formation. The Iverson first began to flow oil after acidation. In some wells they have used as much as 50,000 gallons of acid, pumped in the formation under thousands of pounds of pressure. This is a factor against vegetarian geology, as the oil must have been put in place before, or at the same time, the formation was developed. How could oil and/or gas just seep down there under this great pressure in the formation?

NORTH DAKOTA STRATIGRAPHIC COLUMN

By John P. Bluemle, Sidney B. Anderson, and Clarence G. Carlson

NORTH DAKOTA GEOLOGICAL SURVEY

AGE MILLIONS OF YEARS BEFORE PRESENT	ERA	PERIOD / EPOCH	SEQUENCE	ROCK UNIT GROUP	ROCK UNIT FORMATION	ROCK UNIT MEMBER	MINERAL RESOURCES	ROCK COLUMN (EXPOSED UNITS SHOWN WITH IRREGULAR RIGHT-HAND MARGIN)	MAXIMUM THICKNESS FEET (METRES)
0.01	CENOZOIC	QUATERNARY — Holocene	TEJAS		OAHE		Water, Gravel		50 (15)
		QUATERNARY — Pleistocene	TEJAS	COLEHARBOR	NAMED FORMATIONS — WEST CENTRAL: SNOW SCHOOL, HORSESHOE VALLEY, MEDICINE HILL; EAST: FALCONER, DAHLEN, GARDAR, VANG; RED RIVER VALLEY: SHERACK, POPLAR RIVER, BRENNA, HUOT, WYLIE, RED LAKE FALLS, ST. HILAIRE, MARCOUX (Numerous named and unnamed units)		Gravel, Water, Stone, Peat, Clay		1,000 (300)
2		TERTIARY — Pliocene	TEJAS		(Unnamed Unit)		Gravel, Water		300 (90)
		TERTIARY — Miocene			(Unnamed Unit)				400 (120)
		TERTIARY — Oligocene		WHITE RIVER	BRULE		Clay		150 (45)
54				WHITE RIVER	CHADRON				100 (30)
		TERTIARY — Eocene			GOLDEN VALLEY	CAMELS BUTTE	Clay		215 (65)
					GOLDEN VALLEY	BEAR DEN			
		TERTIARY — Paleocene		FORT UNION	SENTINEL BUTTE ("TONGUE RIVER")		Clinker, Uranium, Coal, Leonardite, Water		650 (200)
				FORT UNION	BULLION CREEK ("TONGUE RIVER")		Water, Stone, Coal, Clinker		650 (200)
				FORT UNION	SLOPE		Coal, Clinker		
				FORT UNION	CANNONBALL		Water		650 (200)
68				FORT UNION	LUDLOW		Clinker, Coal		
					HELL CREEK		Water		500 (150)
					FOX HILLS	COLGATE	Ash		400 (120)
					FOX HILLS	BULLHEAD			
					FOX HILLS	TIMBER LAKE	Water, Stone		
					FOX HILLS	TRAIL CITY			
				MONTANA		ODANAH	Water		

(Continued on page 62)

Era	Age	Period	Sequence	Group	Formation	Member	Fluid	Thickness ft (m)
MESOZOIC		CRETACEOUS	ZUNI		PIERRE	DEGREY		2,300 (700)
						GREGORY		
						PEMBINA		
						GAMMON FERRUGINOUS	Gas	
				COLORADO	NIOBRARA			250 (75)
					CARLILE			400 (120)
					GREENHORN			150 (45)
					BELLE FOURCHE			350 (105)
				DAKOTA	MOWRY			180 (55)
					NEWCASTLE		Water	150 (45)
					SKULL CREEK		Water	140 (40)
					INYAN KARA		Water	450 (135)
	136	JURASSIC			MORRISON			260 (80)
					SWIFT			500 (150)
					RIERDON			100 (30)
					PIPER	BOWES		625 (190)
						FIREMOON		
						TAMPICO		
						KLINE		
						PICARD		
						POE		
						DUNHAM SALT		
	190	TRIASSIC	ABSAROKA		SPEARFISH	SAUDE		750 (225)
	225					PINE SALT	Salt	
		PERMIAN				BELFIELD	Oil	
					MINNEKAHTA			40 (12)
					OPECHE			400 (120)
					BROOM CREEK		Nitrogen	335 (100)
	280	PENNSYLVANIAN		MINNELUSA	AMSDEN			450 (135)
						ALASKA BENCH		
					TYLER		Oil	270 (80)
	320			BIG SNOWY	OTTER			200 (60)
					KIBBEY		Oil	250 (75)
		MISSISSIPPIAN	KASKASKIA	MADISON	CHARLES		Salt	
							Oil	
					MISSION CANYON		Oil	2,000 (600)
					LODGEPOLE		Oil	
	345				BAKKEN		Oil	110 (35)
PALEOZOIC		DEVONIAN			THREE FORKS		Oil	240 (75)
				JEFFERSON	BIRDBEAR		Oil	125 (40)
					DUPEROW		Oil	460 (140)
				MANITOBA	SOURIS RIVER		Oil	350 (105)
					DAWSON BAY		Oil	185 (55)
				ELK POINT	PRAIRIE	MOUNTRAIL		650 (200)
						BELLE PLAINE		
						ESTERHAZY		
					WINNIPEGOSIS		Oil	400 (120)

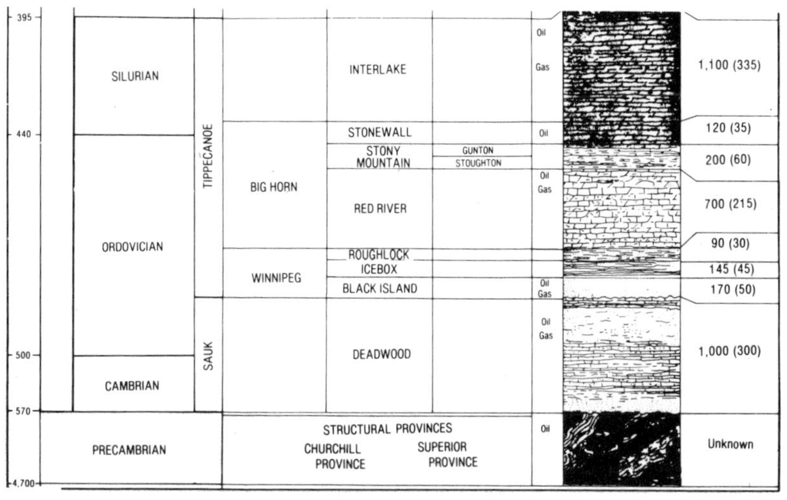

Precambrian may be 16,000 feet deep in the Williston Basin.

The North Dakota Stratigraphic Column shown here is worldwide in scope. In discussing geology of the Grand Canyon a geology book says, *"The Pre-Devonian Unconformity erosion surfaces may be the time equivalents of great numbers of strata elsewhere in the geological column. Thus, between rocks of the Cambrian and Devonian age in eastern Grand Canyon is an unconformity involving no recognizable discordance in stratification, yet, marked by many irregularities or by erosion and representing a hiatus of considerable magnitude. The time involved in this hiatus probably spans all of Late Cambrian, Ordovician, Silurian and Early to Middle Devonian time."*

Could this complex unconformity conglomerate lithology develop by simple erosion, as we know erosion to be today? Elemental distribution will rule out erosion as a developer of this worldwide geological bed of the world. Only annular process could supply materials for these formations over millions of years. Again, elemental segregation must be done before and during placement, not after. Metamorphism has little to do with these geological structures. Many microfossils found here are found over most of the earth in similar strata. Hiatus, a break where a part is missing or lost (gap), is not erosion.

SALT WATER — TEMPERATURE

Some other factors I cannot relate to vegetarian geology, oil is seldom found with fresh water, always salt and salt water. This environment would not be conducive to vegetation. Bottom hole temperature and pressure of oil wells also tell us something difficult to relate to vegetarian geology. I have at hand several drill stem tests of oil wells we once drilled north of Tioga, North Dakota, and will quote from them. Johnston Testers, of Houston, Texas, has this well with a bottom hole temperature of 201 "F", and an initial hydrostatic bottom hole pressure of 4,756 psi. Some wells would have higher pressure, some lower. I was at a gas well in Texas having more than 10,000 psi, wellhead shut-in pressure. The North Dakota well here had a final shut-in pressure of 1,699 psi. Temperature of different wells and depths will vary, some much higher. Just for general information, another bottomed out at 8,250 feet. We used 19 bits, with mud pressure from 700 to 1,200 psi. There was hard drilling at 7,971, limestone, salt and anhydrite, requiring 24 hours to drill 149 feet with 60,000 pounds of weight on the bit, rotary rpm at 90, mud pressure at 1,100 psi, mud weight 10.8 per gallon. Wellhead and flow pressure will always be much less than bottom hole pressure because of hydrostatic fluid weight against formation pressure. With 4,000 psi formation pressure and 4,600 psi hydrostatic, the well cannot flow or blow out as hydrostation pressure is greater.

Again, is it possible to fit vegetarian geology into this complex difference of heat, pressure, densification, salt, all endless variations? Then consider the many types and kinds of oil in one area alone, and also the differences in gas and its content, sulfur, etc. Some gas is not even flammable, one is high in condensates and fluids and another is not. There just is no way vegetarian geology could make all of the carbons of the earth. Consider just one well, say anyplace, like North Dakota. From surface to 15,000 feet we will drill through many formations, some hard, some softer and perhaps several geological ages productive of oil and gas or both, all with a different bottom hole temperature, different pressure, and different gravity of oil and other varying qualities, separated by impermeable sections hard to drill, or they would not be locked where they are.

How could vegetation or animals distill such a fantastic sedimentation of layers down to 50,000 feet in the earth? Life forms require a fairly even environment, temperature and other factors to live, plus sunlight. Temperature alone would rule this out. Would you say the bottom hole temperature at 8,000 feet down of over 200 "F" is the last temperature, and it was even hotter hundreds of years before? Would you care to suggest it was cooler, enough so life could live prolifically, and it later got hotter, after all of this life passed off the scene? Which position will you take? Has to be one or the other. It would seem to me, the last temperature is the one from our test tool gauge. If it changed at all, it was from hotter to what we find now. The answer seems to be earth's igneous, and aqueous strata have largely fallen into place as we find them today, as a giant wreck of slowly declining annular matter. This is now thought to be true by many, as Isaac N. Vail suggested and taught years ago. Annular-canopy has been found in other planets of the universe, so it is not new to us.

The 6:00 p.m. news from channel 8, Grand Forks, North Dakota, May 30, 1980, had a very interesting news item from the University of North Dakota, which said in part, *"Ash falling in North Dakota from Mount St. Helens matched favorably with much older samples taken from western Montana, suggesting then, North Dakota coal beds fell in some 65 million years ago."* This would seem to be very reasonable, giving support to an earthly annular system, and that coal was part of this system and not put there by accumulation of vegetation over millions of years.

An earth annular system does not necessarily fall as a sudden titantic world-collision, but as a continuous shower of dust, meteoric matter, water and all of the materials of the earth's crust, through the "Ages." Two hundred years ago, Kant (German philosopher, 1724-1804) suggested the fall of waters from an earth-ring may have caused the Deluge of Noah. The carbon, hydrocarbon and carboniferous age must have been thousands of years in developing, perhaps millions if you like, under varying conditions, to build all of the geological beds, containing such variety of minerals, oils, gases, coals, predating all organic life and life kinds, as they are mainly carbon, and could not exist without carbon before them in chronology. This age would be so overpowering in magnitude, no human could ever fully explain it or know exactly in detail how it came about. In fact, it is not necessary to know such detail, to know vegetarian geology did not cause this age. There must have been metallic vapors in rings at times, falling in their turn to add to the earth's rich mineral deposits. This would account for stupendous beds so characteristically different from other strata. It is obvious the carbonaceous age lasted so long it came up into the time when life began to appear on earth, accounting for some fossilization in the carbon beds, as foreign matter, as well as in other strata, which fossilization did not cause either.

National Geographic is a most interesting magazine because of its wide coverage of material. On page 680 of the November, 1979, issue, the story on ocean life is outstanding. Strange World Without Sun, is an interesting article about sea life. On page 688, I found a very interesting comment, *"Sea water seeps into cracks, becomes superheated, and picks up minerals from the crustal rocks. As solu-*

tions of hydrogen sulfide rise to the ocean floor, bacteria metabolize it and multiply, creating the primary food source for higher organisms. The discovery of an ecosystem based on chemical synthesis overturns the conventional idea that sunlight is always the main source of energy and life.'' My question is, and perhaps someone more qualified can answer it, is this: Has there really been found a new source of life and energy in this ecosystem, or is it just latent energy and life left behind by the same source, the sun? How else did it get there? Please answer my question. Did not original creative processes produce it?

Where does the energy for an earthquake come from, or the eruption of Mount St. Helens? Is it not latent energy from an original source, spending itself at a later time? When the flood water fell on the crust causing faults, so the earth is slipping back to neutral positions, not with new energy, but latent unspent energy of an original source caused by the flood. Could this be true of the new found ecosystem?

GLACIAL EPOCHS

The entire earth's geology is fluvial, not glacial. The Ice Age is mythical, not factual, just as vegetarian geology is a wrong concept. The time seems long overdue for the world to change this concept, not that it will make any more oil or change geology. I will not quarrel with anyone on the subject. Confucious once said, *"He who strikes the first blow, admits defeat."* 551-479 B.C. Wrong concepts seem to hang on for a long time, even when finally disproved. Ptolemy, astronomer of Alexandria, set forth in 150 A.D. that the earth was the center of the solar system. Others claimed the earth was flat. The one held sway for 15 centuries. Galileo upheld the Copernicus theory, that the earth moved around the sun. Long before the Copernicus theory was proven to be right, the great universities and public opinion were against it as well as the Church, so it was forced out and he (Galileo) was persecuted and banished. According to one authority, Giordano Bruno (1548?-1600), philosopher, was burned at the stake for teaching the world is round, among other reasons as heresay. So it is with vegetarian geology and the Ice Age Theory, it takes time to make a correction. The power of propaganda over truth is appalling.

Most everyone is familiar with the "Great Ice Age" theory or the "Glacial Epochs", as taught by modern schools of science. It would require a book to go into detail on this subject. I will use information gathered from John Tyndall, Dr. J. Croll, Vail, and others. The question is, what must have been the source of those snows that built a mighty continental ice-cap over the Northern Hemisphere during the last glacial epoch?

We are not talking of small local glaciers on a few mountain peaks of today. Stated briefly, Tyndall said, *"Snows, to be formed, require the expenditure of solar energy, and the greater the amount of snows, the greater the energy required."* In short, it requires a great expenditure of solar heat to secure the formation of vapors, before snow can possibly accumulate. The very energy required to melt the glaciers, is the same that would necessarily augment and perpetuate them. We cannot expect the earth to become covered with snows by cooling it, and stopping the formation of aqueous vapor. The glacial theory above must be rejected because it does not present a natural scheme of causation and sequence. It is antagonistic to natural law. Can it be possible that during the glaciation of a hemisphere, that hemisphere can be both warm enough to vaporize the aqueous elements, and cold enough at the same time to build an ice-continent, embracing millions of square miles?

Is it not a fact that they do not now accumulate in any land. Neither in temperate latitudes nor in frigid climes can glaciers indefinitely accumulate by evaporation and congelation. Men of science must not conclude that glaciers always accumulate by the puny process that now builds ice heaps in a mountain valley. Why such a comprehensive and universal change, so sudden and appalling, which left a tropical and semitropical plant and animal world, now known to have existed right under the glacier covered hemisphere? Therefore, they did not accumulate as glaciers do now. This is the great enigma that puzzles so many. The earth does not now have any source from which such a mass of ice could be supplied. There must be an answer. I am convinced, it had to be the grand and all-competent source of tellurio-cosmic snows in the earth's annular system. Perhaps during, and simultaneous with, the Deluge or Flood. The foundation of the glaciation of the planet was laid in the igneous era. The present process of vaporization and congelation under solar influence, could not possibly have laid the great ice beds of the world. As explained above, the two actions work counter to each other for a balance, and they remain just about where they were laid.

We must look at polar ice, as a cap on a man's head. If, as claimed by modern science for many years, glacier ice slid down from all sides over continents, without an equal replacement process, not now existing, why shouldn't there be much less ice in polar regions? Because of the life we know existed under the ice beds, they must have been laid by a one time single process, and have remained until now, because of a balance in the energy force. If we add the necessary energy (solar), they will not form. If we withdraw the source of energy (solar), ice beds cannot form. We now have the answer to the enigma and, as always, it is balance, after cataclysmic causation and placement.

The philosophy of "uniformitarianism geology" is almost universally accepted by learned men. Would you care to say that the twisted, upheaved mass of the Himalayas, the Andes, the chasms of the Alps, the Canadian Rockies, and the mountains of North America, with beds upended, creating the earth's topography as it now is, were caused by slow creeping glaciers and ice sheets? Or would you rather say it was sudden and catastrophic, as I claim? Uniformitarian theory demands an immensity of time. Sir Charles Lyell (1797-1875) tried in vain to link and blend fossils from one stratum to the next. Where are fossil links of life forms between various strata? Prominent evolutionist, Lecomte du Nouy, says, *"There aren't any, there are no transitional forms, it is practically impossible to authentically connect a new group with an ancient one."*

It has never been easy to go against all of the Great Thinkers of the World. Imperfect men do not easily admit error. Pride is more important to them than fact. This accounts for false and wrong concepts being accepted for so long. Great institutions are sometimes built on wrong concepts. This makes it even more difficult to correct them. Pressure for the status quo is very strong on these subjects. Mass public opinion may be manufactured like a pair of shoes, by a mass media. Great educational and religious organizations add to the problem. Who would dare to disagree or oppose them? Fact and truth are squelched as they are too upsetting to things as they are. Save face at any cost.

This is not a Bible based thesis, though I will make some reference to it even though it will not change anything and everything will be just as it is, regardless of what anyone may think or say. For those who do not accept scriptures as fact, that is okay, nothing will change anyway. What I would like to show is the origin of life by science and paleontology is not correct. Nothing put forth by science so far has ever disproved the account of Genesis. The 6 days with God resting on the seventh are not 24 hour days, but creative time periods perhaps several thousand years in length, set out to prepare the earth for man to live on it. God, as anyone would know, does not get physically tired. He only desisted from a certain work on the seventh day.

The Hebrew word "day" is not always from a word meaning 24 hours. We sometimes use the word that way, such as in the "day" of Roman Empire, etc. It is a time period only when used that way. The origin of the earth could be billions of years ago then, and not conflict with the Genesis account. Genesis 1:3 says light came to be. Light had always existed, so now the atmosphere of the earth was first cleared of thick darkness and light came direct to the earth's surface. As science will show, plants appeared to make more oxygen for animals that followed, some disappearing before man came on the earth to breathe oxygen. Also, Genesis 1:3-9 says there was water in the earth, on the earth, and in the expanse above the earth. Falling rain then was not needed, so the first rain was years later at the Flood of Noah's time. After that Genesis 8:22 came to be said, and after that there were seasons, winter and summer, never known before. None of this is complicated or hard to understand. None of the Genesis account disagrees with true science. Nor do any of the great scientists disprove it. They just disagree and in their vain effort to prove they are right, only prove further the correctness of Genesis. The rainbow did not exist before the flood, nor does the Bible say it did. There was no scientific basis for the rainbow before the flood, with a water canopy above the earth, with the sunshine coming through the water making the earth more tropical all over. Some may ask, if knowledge of a subject does not change things as they really are, why make an issue of it. *"To propagate and teach one error hides a multitude of truth. Error taught in the name of science is a pernicious falsehood."* (VAIL)

WATER — TOTAL VOLUME

Perhaps some thought should be given to water to prove the annular theory is correct, and vegetarian geology is not a true concept. Isaac N. Vail, I have quoted before, who did his work before 1912, said, *"All Worlds Made Alike."* Much was said about the vast amounts of carbon, oil, gas, coal, both recoverable and nonrecoverable in the earth. What then about water? Is all of the water in the oceans? No, vast oceans of water are deep in the earth. How did it get there is the question? Vail said it was there, as well as Dana, long before any rotary bits were probing the earth for thousands of feet and finding shallow fresh water and deeper salt water. Texaco Oil Company has an ad, page 15, Time Magazine, August 21, 1978, saying, *"We've got to pump up to 5 barrels of expensive water to get you 1 barrel of oil."* This water causes corrosion, most of it is salt water. It is costly to separate from the oil with a treater, which uses fuel and heat. On land oil wells produce millions of barrels of water every day, and it must be reinjected into the earth if there is too much at any one well. On this date above, Texaco was producing 540,000 barrels of oil per day, and over two and one-half million barrels of water. Multiply this by the entire oil industry and we have a real water problem. I have on my desk a small bottle of oil direct from a Burke County, North Dakota, oil well. The salt water has settled out and the contents are ¾ salt water and ¼ oil. Some have less water, we are thankful for, yet I have seen some make many times this amount of water to oil. Salt water may ruin a good oil well at times. How did it get down there so deep in the earth in the same structure as the oil, trapped forever beneath an impermeable barrier?

In harmony with this, a well being drilled by Gulf Oil, when at 8,000 foot depth blew out April 6, 1982 near Grassy Butte, North Dakota. It was the Foley Stewart Federal 2-24-3C in McKenzie County. Estimates of 5,000 barrels of salt water with non-flammable nitrogen gas per hour roared out of the well bore for several days until well control specialists 'Boots & Coots' stopped the flow.

There are oceans of water in the earth. The geologist must deal with this problem in oil production, yet they scarcely ever discuss the subject as how it got there. Some seeps down from surface, water wells, etc., yet we know this other amount did not run down after the geologic ages completed their work, any more than oil did. The salt water sample with oil I have is so salt loaded, I have a few bottles in the garage that do not freeze even at 40 below zero. Dana estimates that the oceans would be 400 feet deeper if these imbibed waters were returned to them. Others have estimated up to 2,500 feet more water and one 8,000 feet deeper oceans, if ground water were all returned to the oceans. If the seas of the earth were never any deeper than now, when did the water get there and how? It is a simple calculation, if they were all to return to oceans, most coastal cities would be covered with water, and we would have even less land area than now.

As Vail suggests, the ocean is now thousands of feet deeper than it was in Devonian times, to say nothing of Cambrian age. Could it be, we have arrived at a foundation of fact, a theory, that no man can shake, which is caused by progressive and successive collapse of great world-canopies of aqueous vapor, a ring system of annular matter, falling throughout geologic times. It seems then, the only answer to the vast deep underground oceans, came about through this system and were placed where they are, when the oil and other matter in that age were laid down, and never were a part of the oceans we know on surface. Notwithstanding, there is some surface seepage, not deep though, as this water we are concerned with.

Next, we would say the Flood added additional water to the present oceans, and now with air currents, etc., there is natural rainfall, not known or needed prior to the Flood. I apply this remark to the period after man and life came upon the earth though, and not to all of the ages. Most water produced with deep oil wells has no useful value, other than reinjection to help re-energize one or more of the oil reservoirs being produced of oil. The water is either brackish and very salty or both. It will damage crops and ruin land. A certain amount may evaporate, though the only good way to get rid of it is to reinject into the earth at a safe level where it came from. This is another reason to rule out vegetarian geology.

The geological ages mentioned here had a very poor environment for any form of life. As Vail said, *"The aqueous strata began to form as the vapors began to descend, and that the later continued their decline through all geologic time."* Vail proves Kepler's Third Law (a German mathematician, 1571-1630) supports the idea of terrestrial rings. There is so much more to say on this subject, though this will give us some idea of the amount of water the earth contains, and what effect it has, and how it was put in place where it is to this day. See National Geographic — August 1980.

Some laws to consider which are universal are, gravitational, centrifugal, and velocity. Unknown conditions may modify the operation of these laws, yet they are very dependable, and permanent. This is what keeps matter, and the universe, together these billions of years. Someone said, physical fact is as sacred as a moral principle. Legends of the flood, already mentioned is a long subject, and I will not comment on them. Over 20 years ago I came across some equations that had been figured correctly perhaps 80 years ago, before space flights were accomplished. Earth's Annular System, page 22-24, gave these figures, and as I sat in my big chair watching the first man set his foot on the moon, these were the same mathematics the TV Station was giving on space flight. One is, any body moving over 17,000 miles per hour around the earth, would rise to its appropriate orbit, and never until its velocity be diminished to about 17,000 mph, would it return to the earth's surface. As I recall, they gave a speed factor of about 17,500 miles per hour for one earth orbit flight. That was very accurate to predict 80 years ago. Other information and equations are given to prove this theory and how an object could leave its primary, as space flight to the moon does. As long as centrifugal exceeds gravitational force, it must stay up.

CHAPTER 6

KEPLER — UNIVERSAL LAW

With the great success of many space missions such as Pioneer, Viking, Voyager, and others, men of past history are being given honorable mention because of their contribution to knowledge of the universe and science. They are deserving of such respect. Some are: Ptolemy (about 157-170 A.D.) taught the earth was a stationary globe, the planets revolved around. Nicolaus Copernicus (1473-1543) a Polish astronomer, showed the earth rotates on its axis and revolved around the sun. Tycho Brahe (1546-1601) Danish astronomer, believed the Ptolemy theory. Ironically he developed the table, which disproved the Ptolemy theory. Johannes Kepler (1571-1630) was successor to the illustrious Brahe. Kepler was a German astronomer and mathematician. Galileo (1564-1642) was an astronomer, mathematician, and physicist. He improved the telescope, which proved the Copernican theory. He was condemned for heresay by the Inquisition. Also, Sir Isaac Newton, and gravity (1642-1727). The famous equation, $E = mc^2$, by Albert Einstein (1879-1955) a one time Swiss patent clerk.

A man I have already mentioned, professor Isaac N. Vail, who wrote several books and developed the Annular Theory (ring systems on planets), has been proven to be so accurate in his work with physical law, it seems almost uncanny he could prove so many things now confirmed by the space missions. I do not mean to imply he had some mystic or prophetic power. No! He had such an advance knowledge of physical law, from which he would never deviate, as building blocks of facts. I will here give a few of the facts just proven by modern technology, which he taught over 80 years ago, much of which was written before the 1900's.

He taught carbons are not a combination of organic matter. That there would be carbons found on other planets, as well as oxygen, plus large amounts of helium and hydrogen. Oxygen is now thought to be on the Moon Titan, of the Saturn system. Leading news magazines have very good articles on the Voyager mission. Some have made remarks like this: Saturn's rings of ice may duplicate the process that formed the solar system. Also, by studying it, they hope to learn more about the structure and dynamics of the Earth. Too, that Titan may hold the clues for early development of the Earth. Vail had said these same things long ago. Vail died in 1912, yet in his earlier work he described the rings on Saturn about as they are, with the Voyager pictures. Vail said: *"There would be many rings on Saturn, not just a few."* He also said, *"Saturn would have many moons."* Now thought to be over 20. Also, the space between rings once thought to be empty Vail said, *"Would have something in them."* Scientists now estimate 1,000 rings on Saturn. Vail said, *"We are forced to the conclusion that Mars has an atmosphere."* News articles now say that at least two rings on Saturn were found to be eccentric, out of round, an unexpected discovery. What amazed some experts was a braided pattern of debris in the other rings of Saturn. Vail wrote years ago, *"Saturn has what is called the 'square shouldered' aspect. This, I conceive, is really explained by planet's vapors revolving in ellipses, which they necessarily must do. The forces existing differ in degree."*

Vail said, "The earth possessed rings and belts throughout all geologic ages." Too, *"There is an exorable and universal law of planetary evolution."* Two planets, Jupiter and Saturn, giant worlds, undergoing the same stupendous ordeals that ages gone by, our little earth experienced, he says. Vail asks, What kinds of matter constitute those annular and belt systems? *"They are composed of the same material in kind, that now composes the bodies of the planets themselves,"* his answer. He said, *It would be about one billion miles to Saturn, and that Saturn had a thick cloud cover, above normal weather patterns, with many storms therein."* It is now said winds there are 900 miles an hour.

Vail suggested the earth moon had a tugging affect on the rings once existing around the earth. May I suggest, the moons on Saturn, also have such an affect on its rings. In time these rings will descend to the planet, leaving each of Saturn's moons to find new positions, because of attraction loss. They cannot leave Saturn, it is a matter of proximity. A smaller closer moon, could join the primary body, though this is only a guess. Vail agreed, planets and satellites move faster, when traveling closer to their primaries. Again, I will suggest, this too could be a tugging affect.

This seems too little to say about such a great man. However, in this edition of my book, I will include several pages from Vail's book, "The Earth's Annular System," from the chapter, "All Worlds Made Alike", giving his explanation of Kepler's Third Law, in support of earthly ring systems. On one of the "Cosmos Shows", Kepler was given very high regard. Quote from Vail:

Again, it is today a favorite theme of astronomers that, during the igneous era, the earth rotated in a period of only three or four hours. If this be true, the probability that the matter in the primeval atmosphere was whirled into belts or rings is increased from six to eight fold. It seems scarcely needful for me to say, that astronomers came to this conclusion by a legitimate process of philosophic deduction. It must be evident that this rate of rotation would be of great advantage to us in establishing annular conditions; for, almost every schoolboy has learned that if the earth should rotate more than seventeen times as rapidly as it now does, the oceans at the equator would be whirled into space, and made to revolve around it. Then, a rotation in three hours, or eight times as rapidly as at present, would whirl matter already floating in the atmosphere to a greater height and increase annular tendency in the same proportion. However, we will decline to make use of this advantage, and use only that rate of rotation that everyone knows to be correct, vis: one revolution in 24 hours.

Here, then, we have true philosophic data which all men will certainly admit to be fair, and upon which all may proceed to erect the annular theory, and we will endeavor to square every timber in the edifice by one unvarying rule: Philosophic Law. If we succeed with these data to start with, men of science may multiply its certainty by at least twelve, for their own satisfaction. The data then are: a primeval atmosphere admitted on all hands to be 100,000 miles deep, and a known velocity of rotation of once in 24 hours.

With this rate of rotation, we also know that the velocity of any point on the equator of the earth was about 1,000 miles per hour, while the equatorial periphery of the great vaporous atmosphere moved with an actual velocity of more than 25,000 miles per hour. This, the most ordinary mind can determine; but as we are searching for facts that any child who may peruse these pages may understand, I will give the simple calculation here.

If the atmosphere were 100,000 miles deep, and the earth 8,000 miles in diameter approximately, the diameter of the sphere would be 208,000 miles, and the circumference a little more than three times that or about 624,000 the space that any point in the outer boundary of the atmosphere would move through in 24 hours; and, of course, 1/24 of that distance in one hour, or 26,000 miles (I will give 1,000 miles to the other side out of pure liberality).

The simple conclusion drawn from this is, as anyone can see, that a ton of matter at or near the equator of the earth would have a momentum of 1,000 tons, in the rotating mass, while a ton of vapor or any other matter on the peripheral boundary of the atmosphere, would have a moving energy of 25,000 tons.

Suppose the former were laced ten miles above the surface of the earth, and the latter brought down to the same position; the former, with a velocity of 1,000 miles per hour, would immediately fall to the earth, while the latter would rise, and revolve around the earth as a satellite, as can be readily proved by a simple calculation. The mass possessing 25,000 tons of moving energy must lose 8,000 tons of that moving force before it would, or could, reach the earth; for as I have before stated, it is a well-known fact that any body moving around the earth at a rate of more than 17,000 miles per hour, can never fall to its surface, and a ton moving at that rate would possess 17,000 tons of momentum, and it becomes a known fact that if that momentum were increased to 25,000 tons, or a velocity 25,000 miles per hour, it would rise and revolve in its appropriate orbit about the earth, and never until its velocity became diminished to about 17,000 miles per hour could it reach the surface of the earth. Now it could make no difference whether a body be a ton of stone or a ton of aqueous vapor, it would continue to move around the earth so long as the centrifugal exceeded the gravital force. Hence it is evident that upon the data assumed above, of an atmosphere less than half so extensive, as scientists assumed, and with a radial velocity more than six times less than they claim for the mass, the centrifugal force of a vast portion of the aqueous vapors and other matter in the primitive atmosphere was such as to effectually hinder their fall to the earth, as the latter cooled down and the vapors condensed, it is also evident that the matter in the lower regions of the atmosphere would fall on the withdrawal of terrestrial heat, and it is an easy thing to ascertain the line, or height in the atmosphere beneath which all vapors upon condensing would fall, on account of insufficient centrifugal force or moving energy to keep them there, and all vapors beyond which would remain there because of insufficient gravital force to bring them down.

What, then, must have been the condition of those materials that formed the upper and outer stratum of that great atmosphere after the earth became cool and the atmosphere shrank to near its present dimensions, and all the aqueous matter, etc., to the height of 20,000 or 30,000 miles had fallen to the earth? These must have been vast oceans of clouds possessing a velocity that prevented their descent, and which continued to move around the earth; that is, the earth had an annular system. If any criticism can shake this conclusion, there is nothing in law! One would suppose that this is all-sufficient to settle the question forever, that the oceans did not all fall to the earth at the close of the igneous era, but that such as existed when they had not centrifugal force sufficient to retain them on high, did fall; but I will not put this conclusion aside until I have shown still further the impregnable grounds upon which it is based. It is easy to demonstrate by a mathematical calculation that the above depth of atmosphere and rate of rotation are much greater than that which was actually necessary to produce annular formation about the earth.

The analytical expression used by mathematicians to represent the whole force of gravity at the earth's equator is $g + \frac{c^2}{D}$, where g is the visible force of gravity, or the space a body will fall at the equator during the first second of time; c is the chord of an arc over which a revolving body moves in one second, and D the diameter of the orbit of which c, or the arc, is a part, and $\frac{c^2}{D}$ is the centrifugal force or the part of gravity destroyed by rotation, or movement in an orbit. It is evident that the arc c or the space passed over by the moving body in one second, will be practically equal to the chord of the same arc, and I will, therefore, use it as such; that is, as a straight line. Now, as $\frac{c^2}{D}$ is the centrifugal force, and g the gravital or centripetal force, when these forces are equal, and the body neither falls nor rises, but moves on continually in its orbit, $g = \frac{c^2}{D}$.

Now there are 86,164 seconds in one complete rotation of the earth, and the circumference of the earth is D X 3.1416 nearly, and this divided by 86,164, number of seconds in one rotation, gives the length of the arc c, or the distance any point on the equator moves in one second of time; in other words, the rate of motion. But when $g = \frac{c^2}{D}$ it is evident that $gD = c^2$ or $c = \sqrt{g\,D}$, and as often as c, the distance a body moves in one second, is contained in the whole circumference, so many seconds are there in one revolutionl; that is D X 3.1416 divided by c or its equal, $\sqrt{g\,D}$, thus: $\dfrac{D \text{ X } 3.1416}{\sqrt{gD}}$ = number of seconds in one revolution when $g = \frac{c^2}{D}$ or when the earth rotates so rapidly that the centrifugal force on the equator equals gravity. Then we evidently have 5,069 seconds = 1ʰ, 24ᵐ, 29ˢ, or the time in which a ton of matter would have to revolve about the earth just at its surface

$$\frac{D \text{ X } 3.1416}{\sqrt{gD}} = \frac{7925 \times 5280 \times 3.1416*}{\sqrt{16.076 \text{ X } 7925 \text{ X } 5280}} =$$

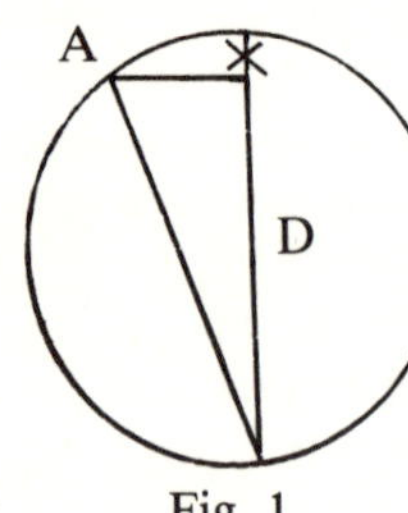

Fig. 1.

*7925 X 5280 = number of feet in the earth's diameter, and 16,076 = g = distance a body falls at the equator during the first second. Let D be the earth's equatorial diameter (7925 miles), and X the versed sine of the arc or distance a point on the equator moves in one second; A X is the chord of the arc, and practically equal to the chord itself, where so small a portion of time is considered.

at the equator, so that it would neither rise nor fall, and when, if its velocity were increased, it would move away from the earth, and in another orbit. Now this velocity is 17 times the present velocity of the earth's rotation, or about 17,000 miles per hour. Hence, we have an absolute demonstration that any body in our present atmosphere or in the great primeval atmosphere, or at any point above the earth, moving at the rate of 25,000 or 20,000 or even 17,500 miles per hour around it, could not fall to its surface! But vast quantities of primeval vapors did not move with this velocity according to our assumed data, which data we have no reason to dispute, and, therefore, we are abundantly justified in the claim that the earth for unknown time was accompanied with an annular system, and the geological record has been misinterpreted, and must be reviewed, and geological theories remodeled.

The foregoing calculations, it might seem, are all-sufficient to establish the fact of annular formation about the primitive earth. But this formation, once effected, demands a permanency of existence, which an immensity of time only can effect. Rings once formed about the earth after the lapse of countless millions of years, cannot collapse in a day. They must lose their momentum with a steadiness as invariable as the flood of ages. It would be as unreasonable to suppose the earth's present satellite would in an hour break lose from its anchorage, and descend to the earth, as to suppose that one of its rings could do the same thing. Then with the primitive earth surrounded with a ring system whose longevity could be counted only by geologic ages, are we for a moment to suppose that the aqueous strata-formation only began after that system had fallen? Is it not more reasonable to suppose that the aqueous strata began to form as the vapors began to descend, and that the latter continued their decline through all geologic time? What is there unreasonable in the claim?

Rings of aqueous vapor, however associated with mineral and metallic matter, must follow in all respects the same laws as a moon or planetary satellite in their motions around their primary. There is a law well known to the mathematical world, called "Kepler's Third Law." Let us bring it into use. By it we can readily demonstrate not only that the primitive distillations, repelled from the fiery sphere, were thrown into a ring-system, but by it we can also readily show how far above the earth's surface they must have revolved about it. This law may be stated thus: The squares of the periodic times of revolving satellites are proportional to the cubes of their mean distances from the primary around which they move. This is of universal application whatever be the shape or constitution of the satellites, as all must know. Then if we take the cube of the radius of the moon's orbit, which is sixty times the equatorial radius of the earth, and divide it by the square of the time of its revolution in seconds, it must be equal to the cube of the orbital radius of a ring of any kind of matter revolving about the earth, divided by the square of the time of its revolution in seconds.

As before stated, the primeval atmosphere in which the matter distilled from the igneous earth existed, and out of which matter all terrestrial rings must have been formed, rotated with the earth; and we have assumed this rotation to be once in twenty-four hours, which the reader will readily grant. Then it must be seen that we have three known terms of a proportion to find the fourth. This fourth term is readily found, and is the actual distance of any terrestrial ring from the earth's center. Put this unknown quantity = R and we will have the following easy calculation. The time of the primitive atmosphere's rotation = 86,164 seconds, moon's time 2,360,608 seconds, and we have the following equation:

$$\frac{R^3}{(86,164)^2} = \frac{60^3}{(2,360,608)^2}$$

Developing and reducing by simple calculation or more readily by logarithms, we will find $R^3 = 279.725264$ and $R = 6.54$ times the equatorial radius of the earth, or the semi-diameter of a ring revolving about the earth once in twenty-four hours. In other words, vapors, of whatever kind, in the primitive atmosphere, at the height or distance of 26,000 miles from the earth's center, or a little more than 22,000 miles from its surface, possessed all the independent energy of a re-volving satellite; and all vapors farther off possessed still greater momentum, and those nearer the earth did not possess the energy of a satellite, and fell to the earth as it cooled down, leaving the more dis-tant matter moving independently about it. Is there anything wrong with this demonstration? Thus "Kepler's Third Law" establishes the truth of the annular theory, or proves itself to be of no value at all!

Then we must see that we need no atmosphere 240,000 miles in depth, nor 100,000, nor even 22,000 miles in order to show that an-nular formation was an absolute necessity in the evolution of the earth. Every mile added to this paltry depth, adds to the certainty of the fact. Did the earth then rotate in 86,164 seconds, or did it rotate in half that time? Every second of diminution adds to the certainty of the fact. How can we escape this conclusion? Thus is rendered plain and irrefutable the claim that the lower part of the great aqueous atmo-sphere, upon contraction and condensation resulting from the loss of terrestrial heat, fell away from the upper part, simply because the lat-ter (like the rim of a great revolving wheel) moved so rapidly it could not descend, but continued to revolve about the earth until it lost so much of its independent inertia, as to permit it to descend, as will be shown in its proper place.

Proceeding thus from the known condition of the primitive earth along a track, every step of which is known, we have by adhering to strict philosophic demand, laid the foundation of a theory that no man can shake. The reader will, from this time, observe that the fabric built upon this foundation, is not an obelisk, but a pyramid, whose successive stages add permanence to the adamantine sills upon which it stands.

Let us look back upon the ground over which we have passed. We see a fiery globe rolling through space with a vast and heavy at-mosphere, rotating so rapidly that its outskirts are unavoidably made to assume such a velocity as to prevent them from falling. The earth was then a glowing sun, or a gleaming star, as analogy seems to prove. But when this earth from "its inmost bosom burned," when its oceans of molten minerals beat upon a seething coast, when its rivers were fluid fire, and its fountains dashing flames, when its "clouds by fiery tempests driven," dropped their steaming floods, an energy potential was stored up in the mighty upper deep — a vast abyss that literally built the aqueous world in after times.

We can easily imagine a world and its atmosphere turning so slowly that the vapors would fall immediately after it cooled down, leaving the heavens clear, and a vast universal ocean washing it. But no such conditions have ever existed on the earth that could bring these things to pass. Every mathematician must know full well, that the rotation of such a mass once in twenty-four hours, would inevitably separate the upper vapors from the lower, leaving the upper far above the atmosphere or terrestrial firmament, obeying the demands of inexorable law. And when investigators recognize this fact, as it stands today demanding a respectful consideration, then, and not till then, will they be able to unlock some of the most perplexing questions of science, which now defy explanation. It is the Philosopher's key to "nature's vast cathedral." I dare not now point out the grand avenues of thought which it opens; but time will make all things visible. I almost said, "all things new," not only physics, but also in metaphysics! All I ask of the reader of these pages is implicit recognition of Law, in this field of labor so near the Great Fountain of Truth. The moment we leave it, we land in shadow and darkness. To propagate and teach *one* error hides a multitude of truths. An error taught in the name of science is a pernicious falsehood. We must, sooner or later, acknowledge the declaration of the missed and lamented Agassiz: *"A physical fact is as sacred as a moral principle;"* for a physical fact ignored sends violated impulses through the nerve centers of society; and their impress is traced in imperishable lines, as by a hand unseen. But one physical fact stands out prominent in the universe, viz: Annular formation is a necessity in the evolution of worlds from their primitive state!!

It must be plain from the foregoing that if geologists had followed the grand train of philosophic events resulting from the igneous fluidity of the burning earth in primitive times, they must have long since concluded that much of the aqueous crust of the earth once existed with the revolving vapors, as infinitesimal particles, or tellurio-cosmic dust, in the ring-system. It must be so; from the very nature of that great sublimation of terrestrial elements, we are forced to this conclusion. And now if men of science will but open their eyes and look, they must see it. Let them follow this conclusion to its legitimate end, and they must see with the keenest regret, the fruitless toil of centuries. No, not fruitless. They have sailed along the very boundary of this field of investigation, and its explored avenues have yielded returns that will aid in the new realm of thought. The glories of a brighter day are dawning in mankind's sky, when all men must see more nearly eye to eye. (End of quote).

ANALOGY OF AN EARTHQUAKE

In considering any action, we must start at the very beginning. Is it simple or complex? Ninety-nine percent of all law is simple, not complicated as science sometimes leads us to believe. For every action there must be an energy source to accomplish the action. Performance is energy. All ideas and theories are worthless until we know what is the causation or source of energy. Where does the energy for an earthquake come from? Does it have its own energy within itself? Does it recharge itself for the next quake? If so, what supplied the mystic energy no one can account for? Could it be gravity or latent energy from an original source? Perhaps our answer is so simple, we have stumbled over it. First, let's ask ourselves, how many possible sources of energy do we have for an earthquake? One, maybe two at the most. Gravity, solar or heat, or both. Which is it?

Perhaps our answer is simple. The earth has millions of faults all around it, not all of which will cause a major earthquake. Even a child would know the millions of faults were not developed simultaneous with the earth's stratification. Faults in the earth cause oil drillers a lot of problems. It is perfectly obvious the earth was bombarded by a cataclysm after it was stratified to cause many of the faults. When the last annular ring around the earth fell in, water and ice caused the earth to buckle and made all of the faults not developed with stratification. This is one source of energy for an earthquake. An earthquake of today does not have its own source of energy, where would it come from? It must be gravity as major faults slip due to stress, seeking an ultimate neutral position.

An earthquake may be compared to a pile of rocks placed where they are by men or natural causes. When we were children we helped Dad haul rocks off the land, placing them in piles. One rock may slip due to stress or disturbance causing the entire pile to fall to a lower level. This may happen several times until an absolute neutral position is found. What was our energy source? The first was solar, as humans eat food from a solar source to get energy to pile the rocks in the first place. We all know the second was gravity causing the rocks to slip and fall to a neutral position where they will remain forever. So it is with an earthquake which has no source of energy within itself. We have no other source of energy but gravity, after the original causation of the faults. Why should we get all tangled up in a lot of scientific theory when the answer is so simple? Due to stress or disturbance, gravity causes earthquakes as these great beds and sills of the earth seek an ultimate resting place, where they will remain forever.

In considering earthquakes, I must make a qualifying remark about stress or disburbance which may start an earthquake. In recent time, there has been a transfer of energy in river silting from all of the continents of the world to the shorelines and nearby ocean beds. In all of the rivers of the world, this could be millions of tons per day. For this reason, I do not give this action hundreds of millions of years as most geologists do. If this were true, all silting processes would have spent themselves and ended long ago with nothing left to silt. Great rivers and silting only developed after the flood which augmented the oceans of the earth. Weather patterns, as we know them, first developed causing rain and snow feeding the rivers and causing silt to move in large amounts. Perhaps this energy transfer by silting may be a stress factor in causing earthquakes, though we do not know. Gravity is the main energy for an earthquake as huge beds of rock seek a neutral position after disturbance.

Conservation of energy is a fixed law of the universe. As currents transport solid matter from one part of the earth to another, pressure increase must be measured by millions of tons in silt accumulation around the earth. This transferred energy is converted into mechanical heat. Perhaps this accounts for many earthquakes and volcanoes near shorelines and where large mountains are, even though they take place far inland as well, perhaps for the same reasons. The Mediterranean Sea is lined with many mountains and has been the scene of many volcanoes. Sediments accumulating in the Mediterranean Sea cannot be borne out to the oceans, so pressure may build there to an appalling degree. The Great City of Alexandria, Egypt, once had an island city now silted contiguous. Many rivers pour into this sea carrying enormous volumes of sediment and is a good example of energy conservation and transfer. The Ganges River carries 900,000 tons of silt each day to the oceans and banks.

Can scientists find any other cause for volcanic eruptions and earthquakes? Volcanoes and earthquakes have no other vent for transferred and transmitted energy. Gravity then, becomes the neutralizer of these forces. This energy must be accounted for, an earthquake does not contain its own energy. It cannot be lost. A map of volcanic regions of the earth will show most major volcanoes are at shorelines where sediment builds up and in mountain areas already having great pressure. Ocean volcanoes in deep water are caused by the same factor by ocean augmentation in later geologic times. The Hawaiian Islands have great ocean pressure on all sides — a natural setting for volcanoes. The oceans of Hawaii are over 18,000 feet deep. The pressure must be great at all times in the deep oceans. Perhaps the same cause may be given to the eruption of Mount St. Helens. A reasonable conclusion is that heat, energy transfer, pressure and expansion cause volcanic eruptions.

It seems evident that all great energy transfers, including ocean augmentation were toward the ocean bottoms. We must conclude, as stated before, once the arch of a continent began to form, it always remained as such. May we never forget the ever present force of gravity in earthquake actions. If we had no other factor to consider, this is one that will always be with us. Without gravity we would not have weight to cause deep down heat and pressure.

Physical law dictates that average continental oscillations would trend higher not lower, which we now have geological proof. To reduce the world ocean levels a single foot would reduce pressure tremendously. Fifty feet would reveal a great continental-shelf. We cannot assign a puny agent to the elevation of continents and to submergence of world coastlines. Remove the cover from a tube of toothpaste and give it a hard squeeze and you duplicate the eruption of a volcano. Such causes and actions are present in the earth in accumulative and diminutive reactions. It would seem the day will come when they will have been all spent.

The annular theory which shows great rings around the earth descended with great force increasing ocean levels causing them to climb upon the shores of the world is the only competent power. Measureless forces would be directed toward the continents which nothing could resist. These grand convolutions have taken place again and again in past ages. We herein see the powerful actions of solar energy in this transfer of energy. Solar energy raises the vapor high; it falls as rain and snow over the earth. Streams feed rivers and rivers feed the oceans and carry sediments along. We see energy employed by sunbeam in raising vapor from the sea to cloud, employed by an energy transfer to a volcano and/or other activity on earth, which must be the same. Why is there evidence of submerged forests near the shorelines of many continents? Answer, ocean augmentation. If there are rock expansions going on in places and contractions in others, expansion would elevate the earth's crust and contractions would lower it. There would then be a slow accumulation or diminuation of this mechanical energy. Law will say if there is a sinking of one region, others must be elevated. When the Great Basin of the central United States was drained, leaving ten thousand lakes in Minnesota and other lakes in the basin, pressures on top of the earth must have been greatly reduced. Pressures elsewhere would have to increase namely the ocean floors.

There is a trend to accept without question popular, long standing theories on any given subject. The Ptolemy Theory stood for fifteen centuries (some say 14). A learning process seems to prevail that new ideas are not accepted until the old one is destroyed. Why this is the case, I do not know. Truth and fact stand regardless of opinion to the contrary. The continental drift theory has been incomprehensible to me from the outset. The more research I do, the less credibility I find in the theory. When science links this theory to plate tectonics and attribute volcanic eruption to plates, many miles thick, deep in the earth moving around lining up holes whereby molten lava then erupts through to the surface of the earth, it staggers the mind. To accept this, we must agree that many tremendous forces are in operation at the same time working counter to each other. Some actions perpetuate, others do not. Consider the millions of valleys, channels and minor corrugations of the earth made by the excavating power of water in a one-time runoff. Today, no water runs through them except when it rains. The great rich fertile silts of the world, such as the Red River of the north were made by a great down-rush of water from annular rings once on the earth. Earthquakes, volcanoes and every action must have a competent force. The origin is Solar. (Psalms 104:1 through 10. Luke 21:11)

SATELLITES

Much has been said about the moon perhaps slowing in its orbit around the earth. To keep satellites in an undeviating orbit around a primary, we must have two forces equal. Gravitational and velocity, which gives us a permanent centrifugal field, which will not change unless we bring in another factor. Some astronomers claim that the moon is moving away from the earth and slowing. How could the moon recede and slow its velocity at the same time? If a sattellite's motion is retarded it will decline toward the central body, until it falls into the central body. Or, if we take the other position, the moon in time will leave the earth and find a new body to claim it or there could be world collision, who knows? Then if this is a constant action, the moon once would have been so near the earth, tides would have covered the continents twice each day. Have we overlooked a factor and is there a simple answer to the problem? Let us say the moon's velocity has been constant and it does take longer to orbit. If the moon is, and has been receding, it is making a longer time orbit with the same motion factor. Let us say in millions of years of earth development, there were annular rings with billions of tons of earth materials in them far out from the earth. Would not this increase the attraction of the central body toward the satellite, pulling it in closer?

If this is true, we do not have to worry about the moon ever being too close nor will it ever leave the earth, as our next argument will be, as the annular rings slowed and declined as Sky-Lab did and fell into the earth, this loss of attraction from the central body, would let our satellite, the moon, move out or recede slowly though, and who knows if it is still receding, in search of an absolute gravitational orbit where it will remain forever. All of this may come about with no velocity change whatsoever. Yes, our time elapse orbit will increase because of distance which is greater. This concept does not violate one physical law.

One thing which has always puzzled me, is where does velocity inertia come from for moving satellites, just in our solar system alone? Their timing is perfect. It is perpetual, why do they not slow down from their initial thrust? These are all located according to unchanging law, at a distance from their primaries measured by their velocities and gravitational force. This brings us back to our original premise, the natural position of all physical law is neutrality.

Much ado is made by scientists who are trying to synchronize the astronomical clock. Who is to say, any piece of a variance must be the dictum of perpetuity? Dr. Stephen Jay Gould ('The Panda's Thumb) says, *"We have known for a long time the earth is slowing down (rotation)."* Science estimates about 1/50,000 second per year. I would agree, the original igneous earth had a faster rotation than now. Could anyone take one year of time and extrapolate this to scale? In driving 200 miles, I may have driven speeds from five to sixty m.p.h., and even stopped for coffee. I know when I left and when I arrived, but which section of time could a guesser take out to use as a gauge to accurately predict this total time span? Perhaps then, the earth is still seeking an absolute rotation, where it will remain for eternity. Scientists feel any disparity gives them an eternal building block. This is not necessarily true. Many theories are given for the earth's postulated deceleration; some are valid; perhaps. Tidal friction is one. Waters of the earth have a tidal bulge toward the moon as the earth rotates. It has been said, the moon's center of mass is displaced toward the earth. In that the moon has sidereal rotation of just over 27 days, and a lunar rotation just over 29 days, may we ask, why its center of mass should be displaced toward the earth? With the same side always toward the earth, perhaps it should be.

Dr. Gould says, *"The moon has been steadily receding from the earth."* It would seem one of two factors, or both, could cause this. An attraction loss, or a circumplanetary orbital momentum speed increase. The attraction loss I have already explained logically. Would the earth's slowing by an estimated 1/50,000 second per year, build enough angular momentum loss, even considering the laws of conservation, to transfer energy or momentum to the moon in such amounts as to cause a rapid lunar recession? Let's think about that one for awhile!

G. H. Darwin, son of Charles, believed the moon had been wrenched from the Pacific Ocean, and he extrapolated its present rate of recession back to determine the time of this convulsive birth. In your wildest delusions, can you imagine the violation of physical law, whereby the moon could be ripped from the earth and not take everything else with it? Was the Pacific Ocean there when this took place? As suggested by scientists in this impossible theory, the moon in its early orbit around the earth would be very close. They have already postulated the recession scale. Anyone would know, if this were true, tidal attraction of the moon would have dragged the oceans of the world across all of the continents of the earth each 12 hours, and washed them flat as a pancake. It must be obvious to everyone, the proximity variance of the moon to the earth has always been within reasonable distance, in accordance with attraction and orbital velocity.

Another factor to consider, was the atmosphere on the earth when the moon was pulled from its crust? If so, how did it escape its drag? If the moon really is a chunk of the earth thrown free from it, did it leave at the equatorial plane? How else could it? Too, the earth is a very watery planet with water even imbedded deep into its crust. If by some miracle upon miracle the moon did escape from the earth's gravity, why then is there no free water on the moon? I do not mean to imply that the moon and planets do not have a material relationship and could not have been parts of each other far back into the creative process. No one yet knows how the earth and the moon first got placed where they are, or when. (See, The Panda's Thumb, Times Vastness.)

Astronomers may take a small piece of phenomena, a microsecond, a microfossil, and use it as a gauge for backward extrapolation, as an absolute for astronomical problems, some of which do not exist. Some paleontologists will take a small fossil shell and postulate the age of the solar system or at least the earth, from this. Coral growth and geochronology, as well as geochronometry have proven very little to me. Here is one for you to ponder; 'observance of corals about 370 million years old.' They could easily be off 100 million years on their age estimate, throwing the entire scale into pandemonium. If the moon became a new satellite of the earth from the earth, in recent geological times, these are only a very few of the unanswered questions needing answers. Sometimes you may wonder if there is any difference between modern science and Greek Mythology.

Does every force and action need inducement after initiation? If all forces of the universe were to trade places with each other, would anything change from what it is? With some forces perhaps the only inducement they will ever need took place at initiation. We are sometimes asked to accept a gauge of such thinness it is almost non-existent to measure accurately the entire universe. Therefore, there must be great error in the measuring stick geophysicists and mathematicians use when they take a millisecond out of a total of billions of years as they claim they can do, then accurately measure a total of billions of years, or anything, or factor.

The moon moves about the earth at an average speed of 2,300 m.p.h. in an elliptical orbit. Was this elliptical orbit established at initiation or was it developed later? Many moons, planets, satellites, travel in elliptical orbits, as the earth does around the sun. Some rings on Saturn are out of round. Did they once have true circularity and become elliptical later? Then an outside force must come into play at a later time. There is room for a lot of speculation here. Fortunately, we know where they are now and their orbital track. No one knows what outside force came into play after initiation, if any. To extrapolate backward seems futile without a true base to measure from.

The intellectuals of the world, the scientists, tell us life is nothing more than the contrivances of nature, with some self-imposed alliances among itself. Survival is manipulated by their own struggle to live, and in the end they lose to extinction. Let us reason they say, so I here quote from Panda's Thumb: *"If God had designed a beautiful machine to reflect his wisdom and power, surely he would not have used a collection of parts generally fashioned for other purposes. Orchids were not made by an ideal engineer, they are jury-rigged from a limited set of components. Thus, they must have evolved from ordinary flowers . . . But ideal design is a lousy argument for evolution, for it mimics the postulated action of an omnipotent creator. Odd arrangements and funny solutions are the proof of evolution — paths that a sensible God would never tread but that a natural process, constrained by history, follow perforce."* Darwin reasoned that self-fertilization is a poor strategy for long-term survival; therefore, the Orchid must have formed an alliance with insects to do the work for them. Now, isn't all of this astounding to say the least! There is an astonishing number of contrivances for life forms which do not even know their own genetic makeup. Science has never proven the brain of a genius is any different from that of an ordinary person. Is a genius born on the periphery of an ancestral boundary, so that he may, by speciation, develop higher mentally? Perhaps the average tradesman has more intelligence.

These are interesting and powerful subjects; yet, you and I are able to make some sound and intelligent decisions. Nor is it easy to go against the well established ideas and beliefs of many people who have been taught to believe a certain doctrine, religious, or science. It is not my intention to be radical, but progressive in thought, as were many men in ancient times who disagreed with accepted theory, sometimes by intimidation, if not fact. With due respect to those who added much knowledge, I have quoted from many.

To propound and stimulate some thought consider some points of view. Why are some things divine and others diabolical? Some people still ask, well who created God? We all know the next question, who created who created God? An endless dilemma. If the ''Big Bang'' theory is true, everything on earth, life down to the smallest flower, or will ever exist, had to be embodied in the first bang. If this be true, we are either riding a Roman candle to oblivion, or ultimately back to the second bang. What will the second bang make? If all of this is nothing more than an accident, without intelligence behind it, could we then say, the highest intellectual thought is only mental dereliction and lunacy, as no matter what we may contrive or improvise, it will have been that way anyway, by accident only, and not by choice? Is then, all conscience life merely delirium tremens? Will we relegate physical law, by iota to extinction? Is there any reason to take the next breath? Think about it, do you have the answer on the above basis?

My main point is this: That vegetarian geology is not a true concept, is a false philosophy and that fossilization adds nothing to, nor did it cause the great carbon content of the earth. Fossils did not cause coal, or oil, any more than they made all of the other rocks and material they are found in. There is no in-between ground, they either did or they did not cause carbon to be. Fossils are a foreign material, when found in coal, oil or rock.

The Ice Age did not form the earth's surface as we know, it is fluvial, from the Flood, and aqueous stratification far back into geologic times. It seems as even the evolutionist would, the earth was formed from great heat, and I add, annular systems of strata build-up throughout geologic times. This accounts for its great varied strata.

Man has not been on the earth for millions of years as some claim, but rather only a few thousand years, as history and archaeology will prove. That all life forms appeared on earth suddenly and remained exactly as they are found, and everything is after its kind, and never has any worthy proof ever been presented by science to prove differently. Just wild, unprovable claims are made without any proof, or real fact. That no real in-between fossils have ever been found, only the false one like Piltdown. Creatures could not evolve over millions of years as how could they breed, or see each other to breed? There appears to be competition among scientific researchers and sometimes this competition to shoot each other's theories down, is greater than the facts needed to do it with. Think of the honor to someone able to prove Einstein wrong.

Continents perhaps do drift and shift around some, but to say they have drifted apart by thousands of miles, I still have my doubts, at least until some real proof is offered. Where would such a force come from, like the idea of glaciers sliding down and covering continents. Who pushed the ice, or the continent? With continents, the law of displacement comes in, we must account for the other side then, and no one has. Perhaps there was a little ice on earth before the deluge, and all shorelines would have moved up because of additional water.

CHAPTER 7

TECTONICS

I will comment on a book, the Geology of the Grand Canyon, put out by the Museum of Northern Arizona, Flagstaff, Arizona. I will recommend this book very highly. The millions of years assigned to geological actions seem too long. Contents cover interesting subjects like: Geology of the Older Rocks of the Upper Precambrian — The younger Precambrian Rocks of the Grand Canyon — Paleozoic Rocks of the Grand Canyon, also fossils of Paleozoic and Mesozoic, among other subjects. It is very interesting.

This work, as do most, makes some admissions and hints toward a very large amount of water in geologic development, which I call the flood. Also to an igneous origin of strata formation. True, the original earth was very igneous, then with centrifugal force, heaved much of its material into space, only to reclaim it later, layer upon layer, as it cooled and slowed, and fell in. On page 12 of this book: *"When microscopically viewed, this rock shows relict hypidiomorphic textures and crystal zoning which strongly suggest an igneous origin."*

Page 13 says, *"Because only small amounts of silicate material can be dissolved in H₂0, very large volumes of water would be required to form the observed volume of pegmatitic rock."* The Flood, as well as antideluvian ages, would provide the needed water, my comment. On page 31, our book says, *"Breccias in the Sixty mile Formation indicate local uplift, rapid erosion and deposition probably in a fluvial environment."* Yes, to this we would have to agree, as again the Flood would cause this.

Figure 11, on page 18 of this book says, *"Folded pegmatitic body cross-cutting only slightly deformed schists. The origin of this relationship is an unsolved puzzle."* They are here speaking of the development of shear zones, which we know sedimentation or erosion could not cause. Yet the buckling of the earth caused by billions of tons of water from the Flood could very easily do this. Macrofossils, microfossils, and fossils, of the Canyon would be very interesting. A microfossil paleontologist would have a big time working here, I am sure of that. Let us leave that for the experts though, as I would next like to comment on Plate Tectonics and the Grand Canyon. Page 106 of this fascinating book says, *"To date, no field evidence has been found that proves recent displacement along the faults of the eastern Grand Canyon. Undoubtedly, however, movements have occurred."* Page 107, *"The driving mechanism for plate motion is poorly understood at present."* It would seem this statement is correct as there is no evidence of plate movement since the flood, nor were there any before the flood, that is of any great amount. Most all plate movement was during the Flood. Minor shifting of the earth and its crust does take place.

At this time I will refer you to National Geographic, June issue, 1980, pages 767-769. This subject is the Canadian Rockies, near Banff, Canada. On plate tectonics, here they say 150 million years ago compression forced the rock to fold and buckle and huge sheets to slide over one another. Having spent time in the Banff area, not that one would need to, I do not see any evidence that this action was slow or prolonged over millions of years. It too, appears to be sudden cataclysmic, like flood water. Also the glaciers in this area did not form until after the flood, large and beautiful as they are, none of them ever were so large they slid down over a large section of central North America. Where would such a force come from to move them, and if they were that large, they would be trapped among the mountain peaks and irregularities?

U.S. News & World Report, June 2, 1980, has a very good article on the eruption of Mount St. Helens. Page 31 discusses violence from the Fire Ring. *"Scientists explain volcanic activity in the area of the U.S. by a new geological theory called plate tectonics, which maintains that the earth's outer face is composed of about a dozen huge plates."*

As these constantly moving plates slip and slide over each other, the immense pressures inside the earth cause volcanoes to erupt through weakened points in the planet's crust. The Pacific's ring of fire is the result of such activity, geologists say.

The almost continuous volcanic activity in the Hawaiian Islands is not caused by such plate movement, however, but by a stationary hot spot in the earth's crust, according to scientists. *"As the massive Pacific plate moves over this hot spot, new volcanic islands are created."* Again, my question is, where does the force or energy come from to move such massive plates around, just as with glaciers the same question may be asked? It seems the stationary theory is the right one. We cannot use both theories at the same time in the same part of the earth. This is so monumental, it has to be one or the other. As to new volcanoes coming through the earth's crust, there does not need to be crust movement. Volcanoes must originate from below, not from some surface cause. Pressure just increases and they come through where they find the least resistance. Energy and heat are still there from the origin of the earth building process in geologic times. A new source is not needed.

What proof is there that these plates are still moving. Perhaps they did move some in geologic times and during the flood; however, when trillions of tons of weight settle into neutral position, it is not easily moved. The one exception would be earthquakes, caused by faults, and here we see a normal process of weight seeking and finding neutral position. This, however, cannot forever perpetuate, after reaching neutrality, energy and force are forever spent in this process. Any further action would require an outside force. Where will it come from? With twelve plates shifting horizontally over each other, it only compounds our reasons for doubt. With billions of tons of water in the ocean, would this not seem to stabilize movement there? Fractional movement is possible, however. It would seem basin design of the oceans with billions of tons of water in them, plus continents on both sides would be less likely to shift than mountains, which appear to be piercement in design. Who will say which is more likely, horizontal or vertical movement, or very little or none of each? Any radical plate movement in deep oil drilling areas would shear off oil well casings or damage them. Also, the oil reservoir would be damaged or lost to this action.

GLOBAL TECTONICS

There is an article in the April, 1971, Reader's Digest on continental drift. This is a very informative article and interesting. I have already commented on plate tectonics; however, this article discusses "global plate tectonics." The study involves the ship Glomar Challenger, and its search of the ocean bottoms. I have followed the research of this ship and its crew with great interest. It is claimed that most astonishing discoveries have been made proving the continents are drifting all over the place. Global-plate tectonics has challenged all traditional views, on a stable planet. Admittedly, some scientists say some of the pieces of the puzzle are missing and not all will agree. It has been compared to Darwinian evolution, or Einstein's law of energy and motion in importance. As many scientists do, the last part of the article gives a glowing description of the seven continents and how all of this came about, assigning when, in millions of years. I give great credit to the Scripps Institute of Oceanography. Because of their work and others I have been able to say, not only is there an oil carbon ring all of the way around the earth, even under the oceans, but also the ocean holds vast unheard of mineral deposits, just as on land. This would include iron and many other valuable minerals.

Let us reflect on the new science of global-tectonics. It is claimed the earth's crust consists of separate plates (ten major ones, subdivided into varying sizes), made of rock 40 to 60 miles thick, which float on the hot viscous mantle beneath them. We must agree the earth once was all viscous due to igneous creative process, and still is hot deep down. True, the continents had to be formed some time. They say 200 million years ago. Who really knows? It would seem these geologic world-building actions have long ago performed their duty and no longer are performing, not to the great extent claimed by global-tectonic science. This plate shift is given as they say, the cause for, and reason for earthquakes, volcanoes, mountains, and for fossils being imbedded on Himalayan peaks. It is said in this article, the Himalayan peaks were once the ocean floor, because of marine fossils found on the top. I disagree with this theory. That is just too radical a change to take place after life was on earth. The answer is simple as I have said before, the flood waters carried the marine life to a much higher elevation than normal. True, the mountains were lower before the flood as the water would push down the ocean floor some and heave the mountains higher. Earth buckling from sudden water weight brings on the earthquakes as explained before. This article has continents drifting around wildly and crashing into each other like car accidents. Where does the force or energy come from to move them? This is not a small task.

Global-plate tectonics could be explained this way from the scientists' point of view: The earth's crust has ten layers of rock many miles thick sliding around and nature is playing a game in an effort to line up holes in these plates, so lava may come through and cause volcanoes to surface. As stated before, they say plates are stationary at the Hawaiian Islands of the Pacific. Yes, we must agree this is true, if they were not stationary, the holes would unline themselves and kill all volcanic action there, which has not happened. Also the islands by now would have been destroyed and fallen into the ocean from the constant plate movement. Now then, they claim the plates are moving around at Mount St. Helens, and other areas of the earth seeking the impossible of lining up more holes. If we take them for their word, then we must ask what is happening to the great thick rock plate under the Pacific Ocean or land mass, if it is stationary in one plate and not in another? Logic demands an answer, is it doubling up under the earth like an accordion? There just is no place for it to go, and we must answer to the law of displacement. Everything must be somewhere.

Would not the law of contiguity say, if a plate 60 miles thick is moving it must do so all of the way around the earth? Considering oceans are as deep as 18,000 feet, to over 35,000 feet in places, how much weight is just a one foot square of water at that depth. It would seem with all of the weight of the oceans, structures must have long ago settled to a permanent position, not much could slide under it. The earth is round like a ball, do we have several plates rotating counter to each other, grinding away? What energizes them? If only one massive plate rotating one way it is even more ridiculous, as they admit it is stationary in places, how could this be? The accountability to physical law rules out global-plate tectonics, it is a **pipe dream.** Dad always told us kids when a problem came up with a motor, and engines, a tractor, or any machine or problem, to just use horsesense. He could fix anything and this was the only rule he ever used, start at the source and anything may be solved. We must apply it to modern technology, too.

For years, many theories have been offered as what the center of the earth is made of or if it is hollow. Some religions claim that is where Hell is. No one really knows though, if I had my druthers, I would say the earth is solid, densified all the way through. The Carnegie Institute of Geophysical Laboratory research have scientists working who feel the earth is solid, and that the heavier material is concentrated toward center. Their research indicates the pressure at the center of the earth is 50 million psi. Who can prove this; yet I would rather accept this than the hollow center theory. If, as is claimed, the earth gets more density as we go down, and pressure increases, which seems reasonable, this is a good argument against global-plate tectonics. Also that ocean floors just did not sink down all around the globe, working against such fantastic force is unthinkable, with a zero displacement area to go to. With ocean augmentation, any sea floors sinking are compensated for by land and mountains moving up, equaling the mass, for a reasonable displacement factor.

Next they argue that fossil and plant life in South America, and Africa are related, and that as a jigsaw puzzle, these lands may fit together. They say fossil plants and fresh water animals, could not have survived a trip of thousands of miles across salt water. None of these arguments are correct. In evolution as they teach, life just forms by itself given the right environment. Who will decide then, which side the life formed on, from their point of view?

Plant and animal life was on earth before man, and continental contiguity was more favorable with much less ocean water, before the flood. Migration would be easier for life forms. Did not the ship Glomar find thousands of miles of islands in the ocean with tropical life in evidence on them, now 5,000 feet under water. This had to be predeluvian. One truth will solve a multitude of problems. Are not all of the major oceans of the world contiguous, causing billions of trillions of tons of continental stability below datum plane?

The earth's formative period has past, and it now exists in a gravitational, centrifugal field, as well as magnetic. Imagine a straight line at sea level from Los Angeles to Tokyo. Midway, how many thousand of feet of water depth would be above this line? The earth rotates about 1,000 mph. Yet the water stays there with only a tide twice each day caused by the attraction of the moon. Billions of years ago when continents were forming, they must have developed where they are today. Until I have more facts and proof I see no reason for any great amount of continental drift in our geologic times. We might compare drift today to the wings and engine pods on a 747 Jetliner. In flight, they are always in constant motion. After it lands and is parked, everything returns to the way it was. Can you imagine what would happen if the force holding the water in a circle we call sea level would suddenly be removed. It is a system hard to change or tamper with.

There has been endless material published by scientists on earthquakes and where they may occur and how big they will be, as well as what Mt. St. Helens, may do or not do. Most of it is wrong as no one knows or has anyone ever been down in the bowels of the earth to see for himself. The Theory is like throwing darts, some are going to hit, mainly by accident. I do not mean to be critical of those who do this research, and there is no disgrace in being wrong. These just are not predictable actions. From past history and data, we know they will happen, but where and when no one knows, only a guess.

MAGNETICS

I must comment on one more subject as it is so fascinating, called the magnetic flip-flop. Scientists at England's Cambridge University came up with a theory that the earth's magnetic poles had reversed polarity many times in geologic times. They feel the earth's iron particles in rock pointed south at times, instead of north. What caused this phenomenon is not yet known. Yet they say geologists have learned to date and read these magnetic reversals like rings in a tree. The last flip-over occurred 700,000 years ago they say. Next they give alot of reasons why they believe this, none of which are provable. They go so far as to claim they know when each reversal occurred, how they arrived at 700,000 years for the last one I do not know. Next they say the theory solves a host of mysteries, none of which I have seen solved. Too much detail to go into now, though.

May I suggest, during the agonies of the earth's birth, what magnetic fields existed, before creative processes settled down to some sort of permanency, and took on identifiable form, no one knows. There could have been total proliferation of magnetics, or for that matter, criss-cross to our current polarization. True axis must first be established. Who knows, and would it change that which now exists if we did know? Powerful forces which were set into motion millions of years ago do not change or give up easily. This may be a dumb question, but for the earth's polarity to change, would not its rotation change too? I am only asking. A simple car battery polarization will not change unless an action is taken to change it. A few battery hook-ups are positive ground, though most are negative. This is a planned thing with an action to do it. Who has the power to reverse the earth's polarization, did it occur by accident or plan? Any generator is just a duplication of the earth's magnetic field in a small way and draws on the same source of electricity. Electricity is not stored, it is generated. A battery is a chemical generator, it does not store electricity in itself. It generates on demand by shorting the poles. To regenerate a battery, electricity is forced back through the battery against polarization, to restore its ability to generate, not to put electricity into the battery.

A simple force called torque is not fully understood. Large airplane propellers have a lot of torque and takeoff until forward speed neutralizes it. Opposite rudder control is needed to offset torque. Some manufacturers of twin engine planes now build one engine to counter rotate to the other, compensatory laws thus neutralizes all torque on the plane. It is still there, but perfectly balanced.

Consider the earth with its northern hemisphere and southern hemisphere. In this part of the earth when we drain our bathtub, the water rotation as it drains is always one way only, never does it changes. In the southern hemisphere it rotates the opposite direction. Then we have high and low pressure systems in weather patterns, each turn opposite to the other and they never will change, in the northern hemisphere. Go to the southern hemisphere and they change direction of rotation with the north. Not to my knowledge does anybody know the answer to why they do as they do, consistently and never change. My reason for mentioning these perpetual laws is to draw attention to the consistency of physical law and how unchanging they are. May I suggest if the earth's rotation could be reversed, these actions mentioned might be reversed, as well as the flip-flop or reverse polarization of the poles. I only suggest this as I do not know. We all do know there is no way the earth will ever reverse rotation, you may depend on that.

DID THE UNIVERSE HAVE A BEGINNING?

The Reader's Digest publishes very interesting subjects, the July, 1980, issue has such an article, condensed from the New York Times Magazine entitled, "Have Astronomers Found God?" Until recent times, many scientists preferred the steady-state theory, which holds the universe had no beginning, and is eternal. The article goes on to show how they are sure now the universe did have a beginning. They call it the big-bang theory; however, the way it came into being is not the issue, it did have a beginning is the issue. After the fact the universe had a beginning, scientists began to ask, "What came before the beginning?" Bolder scientists ask, "Who was the Prime Mover?" One British theorist wrote in his comments, *"The first cause of the universe is left to the reader to insert. But our picture is incomplete without Him."* Some astronomers are curiously upset, it seems. The article says, *"I think part of the answer is that scientists cannot bear the thought of a natural phenomenon that cannot be explained. There is a kind of religion in science. It is the religion of a person who believes that every event in the universe can be explained in a rational way as the product of some previous event. This faith it seems is violated by the discovery that the world had a beginning."* They feel the laws of physics are not valid then. The scientist has lost control he feels. What cause produced this effect? Who put matter and energy into the universe? Was it gathered from pre-existing material? Science cannot answer these questions. The scientist's pursuit of the past ends in the moment of Creation.

It must be admitted, some things are incomprehensible, it cannot be any other way. Perhaps universal destruction would be possible by some madman if this were not true. Are we to replace reason with credulity, or has this in some instances already been done? How many times have we driven down a road only to have a sign loom up in front of us saying—dead end. Two words, yet so final, as we turn around in defeat, or happy to know we need to abandon this course. To trace cause and effect backward further into time, for scientists who have had all of their faith in one basket, would seem like a bad dream, and an insurmountable barrier. Perhaps there is a demeaning now, because of the loss of reverentialness once enjoyed by a philosophy, so widely accepted by mankind in general. Not that the average man knew much about the subject, yet it was very popular. It would seem more people now will learn more about the Creation, and that it did have a beginning. If man knew and could answer all of the facts about the origin of the universe and life, wouldn't that make him God, or at least a god of sorts? Isn't it better he cannot know all?

SEDIMENTATION GEOLOGY

Who would have the know how to come up with an equation that would relate to the overkill needed, in vegetarian geology, whereby our primary product is reduced to oil, coal and carbons, after first satisfying the ravenous appetite of all elemental demands on our primary product? Would you care to hazard a guess as to how many billion tons of our primary product would be required to make a billion tons of coal or oil? Considering free oxygen now needed and present for decomposition, there is not much left of our organic skeletal remains to work with. If we could come up with a hypothetical equation or denominator, whereby to multiply our residual material, to arrive at a total for our primary, to build all of the carbon of the earth, through metamorphism, who could believe or prove it, being so horrendously out of proportion with logic?

Wait a minute, I am not through yet! Sedimentation geologists say all the fertile productive soil of the earth is only a layer six inches thick on an average. This is all that stands between us and starvation. Furthermore, it took millions of years for this soil to develop by erosion, silting, and humus process, or whatever. (I have an entire book on this.) Vegetarian geologists also say their process took millions of years of prolific organic life, to produce all coal, oil and carbon on earth. Millions of years ago when topsoil was just beginning to develop, where did this prolific organic life get soil to grow from, to make coal, oil and carbon? By now would it not have used all available fertil soil, and carried it to the depth of the earth as carbon, oil and coal, leaving us to starve for lack of food? Then after more millions of years more topsoil would form, to support more organic life, only to repeat the cycle. This sounds much like the big bang theory or the pulsation theory.

If we are to assign each of these theories long time spans, accumulative end to end, which is first in order, or simultaneous. Which is your choice? If we say simultaneous, it is laughable as each theory feeds on the other. If we say accumulative, which is first in order, what will be the end result? The backers of each theory claim their time span is from now backward in the time clock. That is their choice, so let's hold them to it. Seldom do we have many positions to work from in an original, here we have three, accumulative, assignment in order, or simultaneous. No matter which of these combinations we use in support of vegetarian geology, our argument falls flat on its face. Is not our problem and its solution harboring on the brink of lunacy?

Evidence continually accumulates to show many islands now covered with water which once were not, because of plant growth once on them. Now they say continents are underwater too. These claims were made long ago, before the turn of the century. We have heard much about the "Lost Atlantis", Ignatius Donnelly wrote much evidence on this subject. It is now believed both the Pacific and Atlantic oceans may have submerged continents. Mountain ranges are now proven to exist in the oceans of the world. A very simple question now must be asked, how could such great continents and islands sink without drawing the ocean waters away from the shores of the continents? Yet we know shorelines of the oceans of the earth have moved up, not down. This is the reason for the confusing theory of continental drift. We absolutely must have "Ocean Augmentation." If the oceans of ground water already discussed, were to return to the oceans it would move the shorelines higher, not lower. Facts leave no room for doubt that ocean waters have been augmented, and are fathoms deeper than in Edenic times. There had to be a flood of water engulf the earth in more recent times after plant, animal, and man appeared on the earth.

Further, if as they claim, continents just sank without ocean augmentation, whence the force and energy to transfer billions of tons of the globe, both rock and water. Doesn't these create a stumbling block they can neither climb over or circumvent? Yes, in the law of displacement, we must account for the transfer of material. If all ocean floors were to just sink, the earth being circular, would this not be against the compaction factor, like driving a wedge, expanding the circumference? Is this possible? The annular theory answers the problem with ocean augmentation. The ocean floors did sink down some from additional water weight, with mountains and land mass being forced up some. This gives us a reasonable displacement factor, as well as an energy factor to do the work needed to be done, avoiding the wedge and compaction, not possible toward the center of the primary. Do you know of any force great enough to reduce the volume of this globe already densified, existing in an unchanging centrifugal field?

To start a concept in error, will only multiply the error, as we try to progress toward resolution. Premise after premise must be turned to in order to cover for the original wrong concept, until finally the first position is lost in the maze of unworkable theory. To proliferate will tend to confuse but not cover for the error in concept. To try to segregate at this point will not hide the real fact or truth. To simplify will only magnify the wrong concept, which is what should be done, to make a correction and start on better ground anyway.

In discussing the origin of the universe, the earth, and life itself, there is no way this may be done without bringing in the entire spectrum of related subjects. Anthropology too must be considered along with others in the main body of this thesis. We live in an age when diehard orthodoxy and status quo are being uprooted by new facts and truth, which never were a part of the original concept.

On biology and medicine I will quote some interesting thoughts. Benjamin Franklin once said, *"The patient survives in spite of the doctor."* Oliver Wendell Holmes, M.D., Professor of Medicine at Harvard said, *"I firmly believe that if the whole materia medica could be sunk to the bottom of the sea, it would be all the better for mankind and all the worse for the fishes."* With all respect for the good medical people may do, it is with interest we find one of the first uses found for crude oil, before Edwin Drake drilled the first well, was to bottle it and sell it for medicine. As we all know, raw crude oil will kill in large amounts, and is deadly poison even in small amounts, taken internally. It may still be possible to buy a former patent medicine called Nujol. One author says this was the beginning of the medical drug industry, the bottled Nujol (meaning new oil), raw crude oil. Much of it was sold to the public for years and after Drake drilled the first well, it was much cheaper and easier to get from the source. There is a commercial on TV, perhaps you too may have seen, wherein they say, "Without chemicals, life itself would not be possible." I presume they mean manufactured chemicals. My question is, if that be true, how did we get from where we came from, to Love Canal and other such toxic waste dumps around the earth, if life is not possible without them?

There is an energy crisis, which will not be solved very easy. The subject would require another book or two to discuss. The amount of carbon as I have shown, in the earth is boundless it seems, no one would be able to begin to comprehend the total it is so great, and undiscovered as yet. This, though, is not the answer to the energy problem. If everyone on earth lived with the same energy demand and living standard as Americans do, the atmosphere would soon be destroyed and we would all suffocate. A change is needed. As Dr. Paul Ehrlich said, *"We have 30 years to do this, that is all."* (Tonight Show) Any crisis nowadays, mostly are manmade, become a global crisis situation. As a prominent person once said, *"Satan, the Devil, will yet dance the jig on the ashes of Civilization."* I am an optimist. God has other plans for the earth and mankind. Revelation 11:18

In conclusion, I will quote the last few words from a lecture given by Professor Isaac N. Vail, June the 21st, 1900, at Los Angeles, California, on the subject of the Origin of Petroleum. *"It is amazing to what unnatural conclusions the vegetation theory leads the scientist. The day is coming when some stronger head than this will meet these ironclad exponents of a false philosophy, and show to a laughing world what an illegitimate birth modern geology is. God speed the day!"*

Perhaps Genesis 1:1 will be in order, *"In the beginning God created the heavens and the earth."*

ALLEN LINDERMAN

PAST AND PRESENT MEMBERSHIP TO

North Dakota Oil and Gas Association, member and director
Independent Petroleum Association, member and director
Independent Oil and Landowner Association
American Petroleum Institute
Petroleum Information Service
North Dakota Petroleum Club of America
Museum of Natural History — Manhattan
North Dakota Wheat Producers, Inc.
United States Durum Growers Association
National Flying Farmers Association, 28 years
Aircraft Owners and Pilots Association, 30 years
Pilot for many years
Oil and gas leasing for years and producing, supervised drilling of

dozens of wells
Smithsonian Institute, associate member
National Geographic Society
American School Oriental Research
Biblical Archaeologist, member
Receive more than thirty scientific and news magazines each month
  and papers
Many dozens of books on history, science, and all subjects including
  geology and physical law.

QUALIFICATIONS AND BACKGROUND

Allen Linderman was born on a North Dakota farm, in a family of seven children. All seven were born at home, as were most children in the rurals of the U.S. years ago. We all went to a rural school by horse and buggy or sleigh in winter. There were fifteen children on my mother's side and thirteen on my father's side of the family. Most of my family then were farm people, and still farm much land in this area. My parents fed and clothed seven children on less money per year than my phone, electric and fuel bill is today. Dad could make anything with his hands, given a few tools and a place to work, which he did. My mother used to tell us kids: Many a rose is born to blush unseen, and waste its fragrance on a desert air. Dad would tell us children to pray as though we would die tomorrow, and plan as though we would live forever. Dad was the rose mother mentioned as he could make anything he wished by hand. He taught us everything about tractors, cars, trucks, combines, and the old threshing machine too. We all owe much to our parents for the things they did for us and taught us. The first few years of our life determines what we will do and accomplish and learn.

My mother was an amazing woman in that she fed and clothed seven of us kids in the 1920's and dirty thirties. When I was very young, I remember how she would sometimes stumble and fall. This was because she was developing multiple sclerosis, which is a crippling disease. She spent 35 years in a wheelchair, and died at 79 years old. Medical science did not give her anywhere near that much time. She kept house and cooked right from her wheelchair. She made bushels of oatmeal cookies and fed them to everyone in the community. For more than 20 years she took many vitamins and minerals every day without fail.

To us kids, my grandfather C. B. Linderman always had every new machine first. He farmed sections of land (640 acres) with horses, then steamers and large fuel burning tractors. He had the first combine we ever saw as well as the first rubber wheel tractor. Some people said he would go broke as rubber tires would never work on a tractor. Long before I was born, Grandfather built a large farm home in Wells County, North Dakota, with six bedrooms upstairs, and one down. Also hot and cold running water, as well as electric lights, run by an engine and battery system. The house had hardwood floor throughout.

Grandfather Garland had a large home too, and a large family. My dad's brother Homer, married my mother's sister Sally Garland, so we had some double cousins in the family. We only have one daughter, Nancy, in our family.

Most members of my family are very inventive and progressive. Dad made and invented many things for his use only. No machine was ever made satisfactorily so we would rebuild it. My uncle, Claude Garland, built the first hopper bottom gravel dump box I ever saw, and sold many. He moved to California in 1936, where he later invented a crane which is sold all over the U.S.A. and many foreign countries. Other family members invented things, too.

In 1919, my father Ross, invented a tractor from the engine all through to the wheels. It had a 4 cylinder engine which only required two connecting rods. This was a very advanced tractor for that time. I still have the blueprints, the tractor never was built. It would be nice to make a model of it sometime.

It was my privilege to design, not the first self-propelled windrower, but the first successful, workable, self-propelled windrower. The first model is still stored in my quonset. I went to a company already in the machinery business with my idea and they built three for me. It was a 20 foot cut, first came out in 1953. It has cut thousands of acres of grain and is still in good working order. This windrower became the basic design of most windrowers built to this day. To my knowledge, I was the first to windrow grain up and back in the field, resulting in a great savings in the harvest operation, with this new kind of windrower. It always gives us satisfaction when we are able to contribute improvement and progress to hard work of harvesting. So it is with me as I see this operation being used. The Versatile Company has a 20 foot size self-propelled windrower like this with a shift platform, which will lay the rows side-by-side for a 40 foot windrow.

Before garden tractors were ever heard of, my brother Gene and I built a small tractor with a Maytag engine in it. We had a small plow and took turns running the tractor or plow. The first motorscooter I had ever seen, was built by my younger brother Jack with a washing machine engine. He would ride it to school a mile and a half.

Airplanes always interested us as did any mechanical machine. When I was very young, I decided to learn to fly, later my brother Gene did too, so we bought a plane. The first model had to be hand cranked. We kept it on the farm, where we learned to land on a dime and get change. In winter we put skis on the plane and every field became a landing field. Finally I was flying many kinds of planes up through the twin engine models. A number of my acquaintances are Sr. Captains with major airlines. I do not own a plane now. My flight logbook shows thousands of hours of flying time.

My interest has always been in mechanical or scientific subjects, though I will admit many have not been a blessing to mankind. As a boy, I asked my parents many questions each day, most children do. I am glad they took the time to answer them. Not having alot of schooling, I decided to educate myself. In my case, I liked this better anyway, and was fun. There was nobody to hold me back and I could learn as fast and what I wanted to know, which was everything I could. History and science were favorite subjects. Next came geology, anthropology, and the origion of creation itself. Archaeology research, too, became a deep interest to me.

When the first oil well was drilled in North Dakota, I became very interested, and recall many people saying they would drink all of the oil found in North Dakota. After many wells came in, I got involved in oil drilling in North Dakota. The very successful company, Cardinal Drilling, drilled the first oil well in what is called the Wiley Field, north of Minot, North Dakota, where Highway 83 and 5 join. Cardinal Drilling was a fine company with some nice people in it. The president of the company is still a good friend of mine. Cardinal was sold to Phillips. We bought leases in the area and drilled more than a dozen wells in the area with other oil companies. This included Cardinal, California Company, Johnson, and others. We also drilled with Cardinal in the Newburg area. As of May, 1979, there were 81 wells in the Wiley Field and more now. Other oil fields later were developed in this area.

Next we drilled in Burke County, with Wilhite and others. Then we moved to the NW McGregor area and drilled an offset well to a dry well a major company had drilled some years earlier. This was in Williams County, North Dakota, location of NW/NW Section 18-159-95. This well bottomed-out at 8,200 feet, with good oil test, casing was set and after perforation the well flowed oil at the rate of 200 barrels per day on 11/64 choke. It would have made more; however, that was our daily allowable back in 1960. There was an oversupply of oil then and the production was geared to the daily demand of the buyers, something like wheat acreage on farms. We did not have any partners on this well. We drilled two more wells in the area offsets, and today there are alot more wells in the field we discovered. This was interesting work and I have spent hundreds of hours on drilling rigs. Still love to get out to one and listen to the engines run. We participated in more than 75-100 wells, drilling or royalty.

These three wells were drilled with a drilling rig rented from a company who had a rig stacked in North Dakota without work to do. We had a geologist and drilling superintendent so this was not a problem to drill our own wells. There was a saving in money this way. Being around oil people also meant being around geologists. I learned much from them over the years. Often I would ask oil people why it is when an oil well is drilled dry in an area the people all believe there was oil and they just capped the well and lied about it. Also why many people say oil companies drill many wells which are productive everywhere and just cap them and not produce them. They all would say, this is what people believe and no one can talk them out of it. Wells are never shut in unless there is a good reason, legal, economic, no market, or mechanical problems, or many others. I know a man who had a number of big gas wells drilled and completed, all shut in because he was miles from a pipeline to market. He was in the drilling and producing only, and not pipelines or market, so until some other company built lines, he could not sell gas, though he wanted to very much to get his money back from the wells.

I am not against schools, universities, and education. They do perform a very useful benefit to those learning from them. Everyone should go to the school of their choice. This is not to say education alone is the road to success. Most really successful people I have known did not have a lot of education. I once knew a very successful lawyer who did not go to law school or college. Schools of higher learning sometimes have a tendency to make us a product — a product of conformity. There is nothing wrong with conformity as long as it is based on fact, truth, and high principle. To conform to good is good. When conformity becomes intimidation, regimentation, of thought and progress, then it is not a good environment. It has been said, necessity is the father of invention. To invent without necessity seems to be futile, and a waste. Progress for the sake of progress only, is not progress at all. Progress should at least enhance utility, beauty, and simplify, enrich life not destroy. Much progress now is progress in reverse.

A geologist friend of mine once said: *Man will go the route of the dinosaur.* He gave a hypothetical example based on time. If everyone kept throwing cans and junk out along the roads long enough, we would bury ourselves to our chin in debris. To perpetuate GNP (gross national production) forever is like offering a penny a day doubled for 30 days, or a million dollars, which would you take. The penny a day is more than the million dollars. Five percent GNP per year long enough he said, would bury us alive (Jim Brown).

To qualify in any field of endeavor, we must first understand the very basic simple things about our subject. We cannot learn from the top down, though many try. Learning, as in most things we undertake, is just a process of elimination. The bottom is the place to start, as fixing a troublesome engine. A very simple action called spontaneous combustion, if understood, is the basis for understanding the entire process of how the geologic ages, developed all of the carbon beds of the earth in the igneous ages. Have you ever been told not to leave a pile of oily rags in the corner of the garage or an unused shed, as it may catch fire by spontaneous combustion, and sometimes did, causing an unknown fire? So it was with the igneous world. Hydrocarbon did the job, not vegetation. It had a distillation process we know, but where was its origin? No one knows. Just read an article in Newsweek for June 30, 1980, on the topic of "Primordial Pond Scum", wherein scientists "claim" they have found cells 3½ billion years old, in Australia. Common sense would tell us that it just is not possible to prove such a claim. No one can really prove such a life for cells that far back in time, as common sense will tell us. These are found right near the surface. This is against scientists' erosion figures. All they have is a series of contradictions.

CHAPTER 8
GENERAL INFORMATION ABOUT OIL

Because so many people have asked me general questions about the oil industry and drilling of wells, I will here give some information which will interest you and answer some questions. The oil industry is so complex no one individual will know all the answers or everything about the business. Many wrong ideas and misleading opinions develop by many when facts are not available. Please understand I do not make excuses for the oil people. They are no more honest than any other profession. After all, businesses are made up of people so there is no difference in this area. For some reason dishonest or wrong doing in the oil industry is always thought to be worse than in other enterprises. Why, I do not know. Perhaps it touches more people. Oil people are not better or worse than those, say, in the medical profession (thought to be sacred by some) or in the building industry, or any other. I do not support wrong doing by anyone. The oil industry is the most regulated and taxed of any business. This costs us at the pump.

For years we have all heard about a so-called miracle carburetor, or a number of them which will make our cars go fantastic miles per gallon. Oil companies are thought to buy these gadgets up by the dozens, making the inventor very wealthy. Reflect for a moment on the ramification of such a project over dozens of years. Do you really believe this? There have been many articles lately in your local papers about the State of North Dakota closing down many gas saving gimmicks. Do you think state officials were in on the big deal? Or that they were just protecting us from fraud? There was an article in the Wall Street Journal, July, 1979, headlined: *"High-Mileage Aids, or Gadgets, are Low on Results."* It said, *"These things are coming out of the woodwork, but nobody has shown us anything that does any good,"* says Peter Hutchins, an engineer who tests car mileage, for the Environmental Protection Agency. A good article to read.

All of my life, I have loved any mechanical device ever made, especially cars, tractors, airplanes, or engines. I still remember almost every part of a Model T Ford. The Model T and Model A Ford, had a mixture control the driver could lean his fuel right from the seat. Us kids helped Dad grind valves for other people on these cars many times. We knew, and Dad told them they were cheating the engine on fuel by over leaning the fuel. Gas was only a few cents a gallon then. Any combustion engine over leaned on fuel, will carry back too much flame on the exhaust stroke, and burn out the valves long before they should, or even perhaps ruin the engine. Having flown many planes up through twin engine, most of them have a mixture control, we are trained how to use it. First indication of over leaning is our cylinder head temperature gauge will go above normal, soon ruining the air-cooled engine. At high altitude we must lean as required. Take into consideration an automobile is made to run from sea level to 12,000 feet in the mountains, without any change in carburization. From 40 below zero to 120 degree "F", and humidity from zero to 100 percent. True, machinery could be made to save a lot of fuel throughout this range of conditions, but who would know how to use it? Also some people pull a big trailer, other do not. So the manufacturer must build an engine to meet all of this range as standard. We are stuck with this fact, and the fuel it takes.

A Model T Ford us kids drove weighed about 1,200 pounds. It did not have a water pump or generator, and at best, made 20 mpg of gas. Not bragging a Cadillac over any other car, my 1975 Fleetwood Cadillac weighs 6,000 pounds, with a few people and luggage. On a drive I once decided to see how much mileage I could squeeze out of it.

On a trip, not going against high wind, I drove very careful, touching the gas peddle softly and babied the car all I could, not passing other cars, coming to a slow stop, and starting very easy. Started with a full tank of gas then filling and using miles from the odometer, it made 20 mpg. Most old model car engines as Model T would not have enough power to run the fan, water pump, 70 amp generator, air condition compressor, power steering, plus automatic transmissions and tires 7 inches wide. We see modern cars do produce much more power on the gas they use. On a ton-mile comparison to a Model T Ford, the modern car makes more like 100-150 mpg. This is the miracle car-buretor.

It is true, cars do make better gas mileage than a few years ago, by design and not a miracle carburetor. Most cars were gas hogs because of being forced to put on smog-control equipment. Now we have the catalytic converter, which changes carbon monoxide to another pollutant, though better gas mileage. I am sure much could be done to improve fuel consumption by way of vaporization and many more. There are many ways to save gas I have seen, but they are too complicated and costly. Some small cars now make over 40 miles per gallon. I will qualify my remarks by saying, it is possible that some ideas and inventions have been held off the market which would save gas. Other businesses have been doing such things for years to protect their market. I am not siding with the big car companies or the oil companies. They have their share of problems now, which are ours, too. Because of necessity, the future will reveal many new innovations and developments in energy fields. Planned obsolescence by industry is part of the problem. Friction causes heat and heat is energy; therefore, friction reducing science has already reduced fuel consump-tion, and perhaps new advanced methods may still improve in this area.

Future cars will make many more mpg of gas by design. Many will be very light and transportation only. It is true a gallon of gas or a shovel full of dirt has enough energy in it to destroy an entire city of 50 thousand people, yet there is no way this can be used slowly enough to power a car on this basis. It must be used instantly in that amount or not at all. If the energy in a gallon of gas or dirt could be utilized slow-ly, it would run your car all of your lifetime without adding any more. This, however, just cannot be done.

OIL AND GAS DEVELOPMENT

The first oil men did not use geologists nor engineers. Just or-dinary people did everything. Oil drillers just bought leases or land in fee simple and drilled where they wished. Most early wells were drilled where oil seepage appeared. This was true of Pennsylvania, Kentucky, or California. The first drilled oil well came in August 20, 1859, 69½ feet deep. Colonel Edwin L. Drake (not a real Colonel) drilled the first well. He was a retired railroad conductor. He was hired because he had a free railroad pass. He, in turn, hired a blacksmith by the name of Bill Smith to do the drilling work. As with most new projects never done before, Drake became the laughing stock of the people around Titusville, Penn. Stovepipe was used for casing. The well started one of the wildest stampedes of history. Oil was selling for $20.00 per barrel at the time. Oil was unregulated and a 3,000 barrel a day well came in catching fire, one of many fires, and loss of life was high. Oil went down to $10.00 a barrel, because of too much oil.

It took years to unlock the secrets in a barrel of oil. The oil industry has wasted untold billions of barrels of oil and gas into the air and ground, first by ignorance and then by greed. It has survived price wars, gouging, and corruption. It has grown so fantastically even oil men cannot keep up with it. Price of oil fluctuated to as low as 8c a barrel to $80.00. In the 1880's, big buyers, notably John D. Rockefeller and associates, got a strangle hold on the markets and distribution driving oil down to 8c a barrel. This ruined many oil men. The first use for oil was, lubricant, medicine and illumination.

Before drilling, oil was gathered from seepage, or pits were dug, some dug tunnels into side-hills and could get several barrels a day. First drilling was done by the old Chinese method of a walking ramp or the spring-pole. Later steam power was used for the walking beam action and deeper drilling was possible and less work on men. At 2,000 foot depth, Manila ropes developed so much stretch, walking ramps no longer lifted and dropped the bit. At 5,000 feet the steel cables, used instead of rope, got so heavy they were hard to lift even with steam power. Though some in later years did reach 8,000 feet. Next came rotary drilling, now they can drill to 30,000 feet. Many thought California oil was running out back then as they felt 5,000 feet was all the deeper they could go. Where many shallow wells are close together, a jackline was used so one pumping station could pump up to a dozen wells at once. In recent years, I have seen them still in use. In 1937, Union Oil Company of California drilled the, then, deepest well in the world, 11,302 feet. Early refining was wasteful, and if too much gasoline was produced, it was allowed to evaporate in the air or was burned as there was so little market for it until the automobile got more popular, now the reverse is the case.

Two technical inventions I will mention briefly, which have been a great help in drilling oil wells, are the Electric Log and Perforation

Gun. After an oil casing has been set through the oil producing zone or zones, and cemented, a perforation gun is lowered into the well, when at just the right location to the oil zone or zones, a button is pressed in the service truck and holes are shot through the casing to allow oil to flow into the well bore. These guns may be made up to shoot a number of holes per foot, in as little as a 2-3 foot section into the oil bearing structure, or many feet of pipe may be perforated, 100 or more if so desired, or leave a blank area if desired. Number of shots per foot and sizes are a matter of choice. Two separate oil zones may be perforated at one time.

The Electric Log reads a record of the well structure all the way down to bottom, on a long paper graph. Several methods are used, one may read from resistivity, or rebound from structure. I have a log report from a well we drilled right here. Our geologist ran a Lateral Log, Gamma Ray, and Sonic Log. Water area and oil area may be located by this system, for exact perforation. Both of these operations can be dangerous. A 30 foot loaded gun, when out of the hole, could be accidentally fired with shots going 360 degrees. This will give some information about oil drilling from the Spring Pole, to Rotary Rig. Now there is some air drilling being done.

OIL GUSHERS

Many have asked why we do not have the oil gushers of the old days. New drilling methods keep an oil or gas well under control at all times nowadays. When one does blow out or get out of control, it is an accident. The old cable rigs worked open holes not filled with water or drilling mud and used a bailer to remove cuttings each time they would build up. When they hit a high pressure formation, it just blew out, with no control until it died down. Much oil and gas was wasted. Wells in North Dakota or anywhere could still blow oil up into the air hundreds of feet if it were not for control. One such well was the Harnell No. 1 drilled in the Santa Marie Field of California. The well blew in December 2, 1904, and produced 3,000,000 barrels before it quit. This was the biggest well ever drilled up to then. A bigger one came in March 15, 1910, in the San Joaquin Valley of California. It was drilled by a man who had never drilled anything but dry holes until then. It was not deep, just 2,200 feet, called the Lake View No. 1. The well came in with a roar, destroying the derrick and eventually leaving a big hole in the earth. For months, the well spouted out of control, estimated 125,000 barrels each day with a 24 inch stream of oil going skyward. It was impossible to catch all of the oil in tanks, so valleys and trenches were dammed up to hold some of the oil and save it. The gusher slowed to 90,000 barrels per day then to 50,000, after 18 months the well stopped gushing. The well made 9,000,000 barrels and 5,000,000 were saved. The miracle was, that it did not catch fire. Such a flood of oil hit the market, crude fell to 30 cents a barrel. Such a well got out of control by accident in the Gulf of Mexico last year and poured its oil into the water, doing much damage.

Gas well blowouts in the early days were even more dangerous than oil well blowouts. Often rock and debris coming out of the well, striking metal on the rig would flint its own fire. One early day gas well blowout was in the Santa Fe springs field. This well did have 900 feet of 8 inch surface pipe in the hole, perhaps not cemented in. The drill collar, and 900 feet of pipe were blown out of the hole, the first three minutes destroyed the rig. Sometimes a gas well blowout would bring up so much sand, it would sand up the well from the bottom and kill the well. This was a blessing when it happened, as many just blew until they died, wasting all of the gas, and sometimes catching on fire.

Early day drilling and production methods, and refining, had horrendous waste. Merely skimmed the cream off oil fields, recovering at times only 10% of the reservoir oil. Pump out as much oil as fast as was possible, was thought to recover the most. Now we know a slower rate of yield will produce far more oil in the longer term from the reservoir. Many early day oil men bought land in fee simple as it was cheap, and did not pay royalty to anyone. Also they could do as they pleased and leave destruction behind with junk everywhere on the land. It was a mess. Federal and state regulations finally came in, this has helped keep a more orderly production, and protect the landowners more. Even with modern techniques, much oil is still left in the formation. Oil is lazy, unless something gives it a push, it lies dormant. Secondary recovery method is a technique whereby some fields yield more oil than initial recovery.

Some secondary recovery methods are water injection back into the earth to where the oil is being produced from, or gas injection, or some other methods. This may be done on just one or two wells or more, or an entire oil field with a common reservoir, well defined, may be repressurized as a unit. This is very complicated though, as it involves all of the oil operators and landowners in a given area to cooperate. Perimeter wells may be reversed out for water injection, a royalty base must be determined for the landowner, from the total project oil recovery. This may be very complicated. Some areas may have several oil producing zones, which must be pressurized independently of the others. This will give you some idea how complicated it may get. This all adds cost to our gas at the pump.

Some of the cheapest energy we have is improved recovery in old existing wells, as most of the equipment on them may be used, as well as marketing systems. Tertiary, or enhanced oil recovery methods, will not solve the energy problem. The energy it takes to produce energy is getting so great, and is a negative factor in the overall problem. Also, the high cost to produce energy and the energy to produce energy may equalize someday if not checked.

Other factors in the energy crisis are the demands of the petro-chemical industry, plastics, fertilizer, and hundreds of related products including clothing. Regardless of all of the vast oil deposits I have mentioned, and yet to be found, U.S. oil output has been on a steady decline and may continue for some time. Old wells lose production rates faster than new wells can be drilled, and oil found. Current oil production in the U.S.A. will be difficult to maintain, unless something changes. There has been a steady yearly oil production decline in the United States since 1970, though the trend did reverse in 1981 with a slight increase. Loss of depletion allowance, and new taxes on oil drilled in the U.S. has had a negative effect. Perhaps there is Divine Wisdom in nature hiding its elements and treasures away from the easy reach of man, as so far all he has done is plunder them. British National Oil Corp. is now test drilling in about 4,500 feet of water. As I have said before, there are endless amounts of hydrocarbons and oil deep all over in the earth, the problem is to find them, and pay the cost. Some may never be recoverable.

I must comment on the 800 page study, three years in the making, of the President's Council on Environmental Quality. The Global 2000 Report to the President, Jimmy Carter. By now, you will have read about this report in the leading news magazines and papers as well as TV news, for August, 1980. The panel warns that time is fast running out for averting a global calamity. It warns up to 2 million species of plants and animals may die off in the next two decades. Also, soil salinity by irrigation will increase impairing its ability to grow crops. That carbon dioxide in the atmosphere will rise sharply. The outlook is bleak they say, unless something is done. There are no "quick fixes" they say.

May I say this is not some sudden revelation, as many people and scientists have been predicting this for years, no one seemed to listen. For years I have warned of such a problem, not in the far distant future, but right on our doorstep. Many are those who have written books and articles on this subject. It seems most people feel the other guy should do something about it, as usually is the case. It is not my desire to mislead you, because the earth still has so much oil, gas and carbons that there is not an energy crisis and we do not need to be concerned about fuel in the future. Many of the qualified people I have quoted in my geological theme, have warned about the coming and fast developing crisis. There is a need to save energy no matter how much is still left in the earth. One way to solve the energy crisis would be to allow everyone on earth to have and use all they wanted and we would soon all perish from carbon poisoning, hence no more need of energy.

Then there are those who feel the 1954 Supreme Court decision, upholding the Federal Power Commission's authority to regulate oil and gas prices helped bring on the energy crisis. They say regulatory agencies held the price of oil and gas too low too long at the wellhead, which ultimately reduced the United States oil rig count from a one time high of 3,137 down to about 800. Also U.S. drillers could not compete with cheap foreign crude, if you can imagine foreign crude ever being cheap. (It was though.) On January 7, 1974, Hughes Tool Company set the U.S. rig count at 1,379. One report set U.S. drilling rigs for September, 1980, at 3,069. August, 1980, there are 89 rigs drilling in North Dakota, maybe a record. (December 1980 — 100 rigs.) Peak 1981 — 140 rigs.

Did you know that in the year 1829, near Bartesville, Kentucky, a well being drilled by Dr. John Croghan for salt, hit oil sand and began to flow. Oil gushed from the earth, ran into and spread over the Cumberland River, then caught fire and destroyed trees for forty miles.

August, 1980, some U.S. and North Dakota oil prices were as high as $41.00 per barrel. One oil publication estimates the "windfall profit tax" will cost each major oil company $9.50 for each barrel produced, and independents $4.75, some estimate the cost of finding oil is now over $9.00 per barrel.

SOME MISUNDERSTANDINGS ABOUT OIL

I will base most of my remarks here on North Dakota oil operations, though much of this will apply to other areas too. You will be interested and glad to have this information, as many of you ask these questions. Some feel oil wells and oil people just drill when and where they please. This was true in the early days of oil drilling; however, now oil drilling is under complete control by rules and regulations established by each state, or Federal at times. Regulatory agencies are set up. In North Dakota the Industrial Commission regulates oil and gas with the North Dakota Geological Survey. The Texas Railroad Commission controls oil there. Oil drillers must have a legal lease to drill on, and get a drilling permit from the North Dakota Geological Survey.

There are rules as to where on the land a well may be drilled, and there must be a survey made by a certified surveyor before drilling, as to location on the land. To drill off pattern more than the allowable distance from the survey, must be decided by the Industrial Commissions at a hearing before them. After oil is found, a spacing pattern is determined by the rules for additional wells. At times, there may be several producing zones, each with a different legal spacing pattern.

The information gathered from an oil well is not secret information withheld from the public. After a given time from an oil well, dry or productive, this information is available to anyone free of charge from the Geological Survey. Drillers have the option to declare a well a tight-hole during the drilling, whereby information is held from the competitors, etc. After the time period is past the information is no longer held back.

After an oil well has been drilled so often we hear people saying and beliving the oil people found oil and, for some unknown reason, capped the well and left all the oil down there for someone else to find. This is not correct, the information cannot be legally withheld. This is not to say that in the early days of the oil industry many tricks were used to keep information away from competitors, and for that matter, maybe a well has been plugged even in recent years which should not have been. More often this was a mistake by the people drilling the well, and not something contrived.

We often hear of someone opening up an old well and making it into an oil well. Someone goofed, that is all. Or the price was too cheap for oil to complete what was thought to be a marginal well. An oil well is completely controlled by laws and rules. The law requires every oil well to have surface pipe set, at least 200 feet on down to 600 or more, whatever applies to each area. This may be eleven inch or larger pipe, which is cemented in the earth on the annulus side and left there forever. In the surface pipe, or the long production string set if oil is found, there is some cement required for the pipe. This is done by placing the right amount of cement in the inside of the pipe, placing a rubber plug on the top and pumping to bottom with water pressure, forcing the cement to the bottom and out around the annulus side. (I have an entire book on cementing wells by the American Petroleum Institute.) The surface pipe protects water wells from contamination and interchange, as well as a blowout protector is placed on the top of the surface pipe. The well is then drilled through the surface pipe as far down as desired. If the well is dry, the law requires a cap placed on the surface pipe below ground. This is the law and not a trick to cheat someone out of oil.

Dry oil wells must be plugged according to law with cement used at different levels. Not a solid cement fill all the way down. Why waste cement in areas not needing it. In this case, there isn't any pipe below the surface pipe. Others have said they saw them putting oats or building material in the well, thinking this too was a scheme of some kind. During drilling there is circulation mud and water pumped down the hollow center of the drill pipe, which floats the cuttings from the bit to surface on the annulus side. Sometimes porous sections are encountered and the drilling circulation is lost in the earth. If cuttings do not surface, no new hole is made until recovered circulation is established. I have seen every imaginable product pumped down the well to plug the porous area from grains to insulation materials as well as manufactured products for this purpose. This is a serious problem, not a scheme.

An oil driller may set all of the pipe he wishes to for his own protection after he sets the minimum required for any area. An example, a well being drilled in the Big Hole Basin, Overthrust Belt, western Montana has set 2,600 feet of 13⅓ inch surface pipe. Later 9 5/8 inch pipe will be set to 10,000 feet, and ending up with 7 inch casing to 15,000 feet if they find oil or gas. The cost for such a well is very high.

Slant drilling is much misunderstood, too. Years ago much slant drilling was done illegally. Slant drilling is still being done, though it must be done legally, by a certified engineer. North Dakota now has some slant drilled wells near the Theodore Roosevelt Park. The wells are drilled outside of the park and slanted under the Park, legally though. In North Dakota there is very little bit wandering and to drill straight down is not a problem, though a deviation hold survey is required on each well. Most wells will not be off more than a few degrees, not requiring a correction. This check is easy to do by dropping a test tool down the drill stem pipe before taking the pipe out of the hole. It sinks to bottom taking readings and is recovered at the end of the pipe when removed from the hole.

In checking a well report I have for a North Dakota well, the hole deviation figures are from ¼, ½, ¾ to 1½ degrees, as an average in 8,200 feet of hole. Changing to the new metric system will be very costly, with many mistakes. When on an oil well recently, the driller told me of an oil well being drilled on the metric system. An error in the pipe tally was made with about 1,000 feet too short casing was set. The well was cemented, which ruined the well, a complete loss. Then I read of an oil venture in ocean drilling where a company bid 70 million dollars for a lease, then drilled a number of dry holes on the lease at millions for each well, then abandoned the project. No doubt they had a number of companies in on the venture, or this could ruin a company, especially if it happened a number of times. Compare this to a farmer getting hailed out 3 or 4 times in a row. In the case of oil, we as consumers pay all of the cost at the pump, etc.

In the case of the windfall profit tax, and also wellhead production tax, the landowner, or royalty owner is hit hard by way of tax, next in line is the consumer.

Ocean drilling platforms may drill several wells from one platform, one straight down and slant drill a number of others. If there are several separate oil zones, each may be drilled into with several wells and one platform may have many wells drilled and produced from one. Again the cost to pipe the oil to shore for market is very high. The hazards of oil drilling are endless. First in drilling the well, many things can go wrong costing thousands of dollars per day in rig time not anticipated. Then when oil is found, the risks in completion are endless and a well may be lost as easy as found due to uncontrollable problems. After a well is completed, many problems may come up to keep the well in production, which all cost a large amount of money. Salt problems are sometimes bad, and cost a fortune to keep the well clean and producing. Good wells make a profit, though some wells may never show a profit after years of production, and some will lose money. For those who think oil is a gold mine, they should go into the business or at least invest money in oil and learn for themselves. Yes, money is made in the oil business or it would not be functioning, yet it is not the free gravy train some may think. Every new tax and production cost and expense, we will pay for right at the gas pump, or to heat our home.

Another problem is, many wells drilled in the oceans or remote areas may not be able to deliver oil or gas to a service for years. Companies may have to drill a number of wells to make it pay and it takes time to set up pipe and delivery equipment. Some wells may have a very high gas/oil ratio and production must be restricted, or shut in the well, until gas gathering systems are built. Some gas flaring is allowed but it is controlled. Gas is in great demand now, so line pressures may be so high a compressor must be installed to force gas into the purchaser's line or none may be sold. A Texas gas man told me he had to install a $50,000 compressor to market gas from one well, though one compressor will service more than one well if the owner has them. Then it takes energy to run a compressor, a negation factor.

There is an endless amount of information on these subjects and you could find some very good information on them other than my writing it here. I have many books on these subjects. I only wanted to give a little general information as I know you will appreciate it. Not everyone can be expected to know all about the oil drilling industry. No one person could possibly know everything about the petroleum industry, it takes the know how of 100 years of work and many people to keep it going. New technology comes out all the time, plus geological information keeps accumulating, an endless process.

The very exciting and interesting subject of the history of the oil industry is not the subject of my book, so I have only given you very little here. There are many books on the subject, though I will recommend one you will enjoy: ''The Greatest Gamblers'', by Ruth Sheldon Knowles. This book is an epic of American oil exploration. If you cannot get it at your bookstore, contact the University of Oklahoma Press, 1005 Asp Avenue, Norman, Oklahoma 73019.

These three scenes are west of Cody, Wyoming. The giant upheaved and tilted rock formations, in conglomerate structure were not created by erosion. Erosion is affecting them in our geological times; however, the builder of them was cataclysm after the igneous agency was removed. The energy required to develop these formations was born and preceded them by millions of years. The energy has now been spent with only slow, tedious erosion left to work them into an almost impossible change.

PART II

WILL EMBELISH
AND
COMPLEMENT
THE FIRST
PART

CHAPTER 9
MORE ABOUT GLACIATION
GLACIATION

For compelling reasons I must say more about the Glacier epochs as taught and believed by modern science for years. Why do scientists who discover a certain process or action always say it is endless, and self-perpetuating and goes on forever? After the energy for an action is spent, unless we have a renewing source, the action stops. A plane running out of fuel at high altitude is a good example. Before any action may take place, we must first have the dynamics and energy. Glaciation of a hemisphere must have been caused by thermodynamics in the great annular systems of the planet earth.

Dr. Stephen Gould, who appeared on the Donahue Show with Dr. Carl Sagan, said: *"There have been many ice ages and there will be more as the earth moves closer and farther from the sun."* This wrongly accepted theory was proposed many years ago by Dr. John Croll who had the endorsement of Sir Archibald Geikie (British Geologist 1835-1924). Many theories have been advanced to explain glaciers, few possess a measure of plausibility. To examine the Crollian theory, because the earth in its annual orbit around the sun, is not circular but in an ellipse. It seems he felt the eccentric circle may increase or decrease slowly, and that when the earth was in its farthest position from the sun, accumulates more snows in its polar region during its winters than the heat of summer is able to dissipate, which after ages of accumulation amount to an ice-glacier, covering a hemisphere.

Let's put this to the test of reason! This theory makes regular recurring visitations of the ice-epochs. How then did the prolific tropical life get under the huge ice cap on the earth? It ignores the law laid down by John Tyndall already stated: *"Snows, to be formed, require the expenditure of solar energy, and the greater the amount of the snows, the greater the energy required. To take the earth from the sun, then, robs it of snows and the accumulation of snows. You must increase rather than diminish engine force."* Also as Vail said many times, a continent encased with ice by means of solar evaporation of the oceanic waters (as opposed to annular water) could never again become freed from its fetters; for since it requires a great expenditure of solar heat to secure the formation of vapors (evaporation is refrigeration), before snows can possibly accumulate, it is plain that the glaciers could not melt unless the heat should become greater. But this increased heat would increase evaporation, and increased evaporation means a greater precipitation of snows and an increased glaciers. If a continent should become refrigerated by increased vaporization how could it possibly become free from its grip of ice?

There can be no terrestrial source of continental glaciers. The interior of a continental glacier could not be fed by snows from the oceans without the fraction of all law. It is a well known truth the famous Sir Charles Lyell in his life urged upon geologists the important fact that from the dawn of recognizable changes in the earth, the change in climate, the extermination of species and the change in ocean's level were a triplicity of change, of changes that always remained unbroken in the order of their occurrence. All the shore lines of the world are known to have moved up onto the continents in recent geological times by ocean augmentation of annular waters and snow. It is the only means by which great snow fields could suddenly entomb a living world. Let us consider the mighty wall of ice on the antarctic continent. Antarctic Continent is a mighty field of ice nearly 2,000 miles across. It is an ice continent beyond the reach of snowfalls from vapor and congelation.

From what source came the mighty casements of ice continents, since it is unreasonable to suppose it came from warmer latitudes? Aqueous vapors must fall before the temperature of the atmosphere is reduced to that of the average in the polar worlds. Contrary to what most persons believe, it seldom snows in extreme polar latitudes, it must be mere accident that snowstorms ever occur there. A Dr. Kane in a visit to the pole when 20 degrees from the pole saw sledge-tracks made several years before. Bones of unfortunate explorers, animals, and articles of clothing lie exposed in plain view for years. Snow then seldom falls in extreme polar worlds.

Let us pursue the subject of ocean augmentation as opposed to the old outgrown theory of hemispheric glaciers slowly creeping down over continents for millions of years. The lost continents and islands of the oceans must be considered. How could such great continents sink without drawing the oceanic waters away from the shores of the continents, and thus increasing the pitch of rivers near their outlets? These and other facts leave no doubt oceans have been augmented.

For sake of discussion, let's say great ice sheets did accumulate in upper central United States, as some say up to four miles thick in Minnesota and surrounding areas. First, there never has been a climatic weather pattern to supply such amount of ice. What then could be the source? There is none. Next we must find where such a great source of energy required would come from. Then we need the same amount of energy to remove this great ice body as the energy needed to place it there. Do we have such an energy force available? No, we do not!

Could evaporation and congelation do this even if we had enough solar energy to supply it? Four miles thick ice would be about 21,000 feet high. We may estimate the area we are discussing would be at least 1,000 feet above sea level (datum plane). We now have over 22,000 feet above datum plane. We may reasonably assume we need at least 5,000 feet or more above this ice bed to allow some sort of weather pattern with evaporation and congelation to carry moisture if we are to say the source was the oceans and seas of the world. We now have a figure of 27,000 feet above sea level. Would it be possible for the thin atmosphere at this altitude, to carry enough moisture to build an ice bed 22,000 feet above sea level we call datum plane? No, it does not seem reasonable. If this theory were true, and some of us were there, it would be possible to ski from this area down hill all the way to the North Pole. We do know that much of the polar ice is below sea level. How could such a massive ice flow move uphill and over mountains to arrive in the central United States? Or would you care to say the ice just accumulated where they say it was from snows which have no known source? If we add enough solar energy they will not form, and if we withdraw the needed solar energy they cannot form.

Some questions arise. Was there ever a time in geologic ages when the current ice fields of polar ice had a higher altitude than now, that is above current datum plane? Was there ever a time when there was less ice in polar areas than now? If as we know, the earth at one time, say Edenic times, had much less ocean water, what would this do to our current ice caps? If we call sea level datum plane would this not also need to be lowered considerably? This would leave no place for current ice at the poles to exist. What was the source of both arctic and antarctic fields? Are these huge ice beds less now than when originally placed there? Perhaps the only loss in size we can account for would be melted ice. We do not have a replacement factor for these ice beds today or ever have had in known geological times. On an average, weather patterns today will about replace an equal amount lost each year by melting — if anything there is a small loss, not gain. The great enigma then is, what was our source of snows to place the great ice beds at the poles originally, and leave a living tropical and semi-tropical world under them? Science has never answered these troublesome questions. If we accept as many do, the Crollian-Geikie theory, as many scientists do, we arrive at a dilemma.

An example of total contradiction in modern scientific thought wherein those scientists working in the field of Bacterial microfossils (Archaebacteria) claim the earth had a temperature about as it is today 3.5 billion years ago, as they claim that is about when life began to develop on earth. As any school child would know, if the earth were hot and igneous as it was then, life could not exist or form. They violate all the Laws of Thermodynamics. Then the other group say we have had many very cold ice ages in order to support their ideas. As the old saying goes: *"We cannot have our cake and eat it too."* In Einstein's formula for the equivalence of mass and energy, it works perfectly, whether from mass to energy, or energy to mass. Thermodynamics teaches us that it takes the same amount of energy to break up a molecule as it does to put one together. Would these laws apply in primeval times as well as today? Yes! Yet, at this late date, scientists still put forth the astounding information that mighty ice sheets, thousands of feet thick, pushed down from the far north and leveled our hills and valleys, obliterating rivers and lakes, even though they do not know where the energy came from to move these great ice sheets, nor can they even prove what the source of the snow was to build the ice.

In recent years climatologists, meteorologists, and scientists have been predicting great climatical change for the earth, which may bring about an ice age, or, a warming trend melting polar ice and flooding coastal areas of the earth. Yes, we do know the earth is a dynamic globe with much change possible. It is easy for scientists to say cataclysmic things will happen, but when and where they cannot say. Anyone could make the same claim without worry of reflections in a wrong way. They are just guessing when they say within 100 years to more than a thousand, or even into the millions this or that may happen. Scientists now have some very sophisticated instruments, yet they cannot answer some very fundamental questions. They never have answered one profound question; how did polar ice caps get on the poles? Have you ever given thought to how much snow it would require to build hemispheric ice caps such as Antarctica?

Some expert scientists and climatologists have given estimates of over seven hundred feet of snowfall per year for a thousand years would be required to build polar ice fields. Now then, the energy for this monumental task is solar. To get enough evaporation and congelation from the seas and oceans of the world, to supply such a fund of snows, the oceans would need to be brought to a boiling point. This, as you know, will not build ice beds of such size. If we take the other position and cool the earth to form ice fields by taking the earth away from the sun (solar) there will be little evaporation and ice caps cannot form for lack of snows.

Men of science asked what is behind the so-called weather shift on earth? Many expert theories are postulated and there is no consensus among them. Some climatologists say the earth may warm up a few degrees and melt ice on the poles. The absolutely opposite would be the case. The additional heat (solar energy) will increase the ice pack on earth. Another theory has the earth in a cycle of thousands of years wherein the earth orbits farther from the sun and we will have another full-blown ice age with glaciers invading from the north. This has to be the most ridiculous hypothesis of all. It would take a vast amount of energy to build such an ice pack. If planet earth is moved away from its source of energy (solar) no snows or ice can form. This has been explained before in my book. We must look elsewhere for the answer which is annular theory.

REVERSE ORDER BY SCIENTISTS

Throughout history, men of science have had things in reverse order all too often. Rather than admit error they try to cover over with another far out theory. These two theories just mentioned offered by learned men, all from the great universities, are as wrong as the Ptolemaic theory, which was almost impossible to unseat. To cool the earth will not form ice. Cataclysmic prediction has become a fad. Starting in the early 1970's astronomers began to predict great disasters and earthquakes particularly for California in the early 1980's with the big one by March 10th, 1982. This all was based on the ''alignment of the planets'' in a straight line behind the sun. Because these ideas are against physical law, I have never put any faith in them. I am sure March 10, 1982 will come and go without disaster to California or any other place on earth. If there are earthquakes in California on this date, it will have nothing to do with the reason given by science. The clairvoyant and religious predicters have had a heyday in predicting disasters, particularly the end of the earth (not world) all to no avail. Perhaps some feel if enough diasterous predictions are made, some will be right by accident and they will get the credit. I do not mean to minimize natural disasters such as earthquakes, floods, or volcanoes. We will continue to have some though most are not pre-dictable, when and where nobody knows. Many disasters today are man-made and could be avoided if it were not for greed. (I am not referring to Armageddon.)   Above paragraph was written before March, 1982.

In the annular theory (rings on planets and the earth) which I ac-cept, all of the unanswered questions fall right into place. The stupen-dous polar ice caps were placed where they are cataclysmically by a great fund of snow contained in telluric rings of vapor in the polar regions simultaneously with water and ocean augmentation by a great downfall of water. No other theory answers these questions.

In this proven theory we do not account for energy to remove the glaciers, as they are still right where they fell. The energy to place them there was developed and contained in the original hot igneous earth as it cast rings of material into space. The little understood prime mover, gravity, brought all of the rings back down to earth again, as they cooled and slowed some. Perhaps these simple proven laws are just too simple for the modern day scientist and geologist to accept. I repeat, Einstein and Capernicus, and others, were questioned and doubted by learned men, yet these same kinds of men accepted Ptolemy (earth the center of the solar system) and Aristotle, who set science back for about 2,000 years (Artistotle 384-322 B.C.). Galileo and others destroyed the erroneous concepts of Aristotle, and much progress was made. One wrong concept taught by Aristotle was, a heavy object would fall faster than a lighter one. Aristotle studied philosophy with Plato and became the teacher of Alexander the Great. It would seem Alexander did not follow everything they taught him, as he made great findings in science, with one of his men giving the almost exact circumference of the earth.

Even today in a highly educated and technically scientific world we live in, the most appalling thing I can think of, is the apparent triumph of propaganda over the truth. As Professor Vail once said: *"Truth must always struggle to the light."* In 1874, Vail published a volume the Great Deluge of Noah and the down-rush of the last annular waters. It is still correct. At that time, he received some flattering commendations and on the other hand some of the most excoriating denunciatory criticisms for "treading on holy ground."

Many geologists claim because of ice and glaciers, rivers once ran in opposite directions. What then about the great fresh-water beds? If the rivers ran in the opposite direction analogy would show that these great fresh-water beds could not have been such as they are — so exclusively fresh-water deposits. It seems like reversing natural tendencies to conclude otherwise. What, then, deposited the tertiary beds of Siberia and North America? If this claim be true, what are now the continents were then the oceans. But the evidence is accumulative that an elevated arch of land, always remained an arch; and we have no evidence that such ever became trough of the ocean. Yes, overthrust belts are easily accounted for in annular geology. In the evolution of continents, once the arch of land developed, it always remained as such, and leaves no room for the continental drift theory. If continents were ever to drift, why wouldn't the force causing the great overthrusts move the continents?

One more consideration which will not support the old-school glacier geologist is the simple comparison to a large river system over the earth. As anyone knows a river cannot flow for hundreds of years without a renewable source of water. We all known that rainfall and snow caused by weather patterns which are solar, renew this supply. Water power as it is called, by generating electricity or running the old mill, is not water power at all, it is solar energy. Now, if as claimed, large slow creeping ice sheets and glaciers were sliding down over continents for thousands of years and slowly melting, without a replacement factor, why are they still there? During drought a river dries up. Why then are the polar ice caps still there? As we just proved, to cool the earth to form icebergs and glaciers will not work, as this environment now exists at the extreme poles and we have almost no snowfall at all. If we increase the solar energy and heat up the earth they cannot form. We must look elsewhere for the answer. The mighty downrush of snows and water came from the earth's annular system. The "eternal frost" that took possession of the region, inhabited by plant, animal and, of course, was the "cause which destroyed them." Those tellurio-cosmic snows took possession of the entire polar world.

WESTERN OVERTHRUST BELT

Because of the great interest in the western overthrust belt by geologists and scientists and oil drillers, I must give a short quote from Vail written many years ago. May I call attention to the fact many geologists a short time ago said no one should drill for oil in this area. Now it is believed to be a great oil and gas reserve where drilling will be done many years from now. What force thrust these great beds of rock and earth up over each other? Quote from Vail: *"From this again comes the necessary conclusion that the oceanic waters were greatly increased in volume, and we must therefore expect as a legitimate and necessary consequence a system of upheavals and strata-folding correspondingly stupendous and grand. Must I point my brother geologists to the well known facts that support this latter conclusion? Need I tell them that the continents grown more stable with time would resist oceanic pressures longer, but when they began to move would move with grander strides? Need I point to the greater convolutions of the earth's crust, known to have been formed immediately after the carboniferous beds were laid down?"* This explanation is sound and gives accountability to the geology of the overthrust belt as well as many other puzzling geological questions. Many geology books I have read claim these upheavals and overthrusts as far in as Banff, Canada, took millions of years of slow grinding time. I must agree with Professor Vail's theory. It is entirely in harmony with Law. It was caused cataclysmically and in a short time span, perhaps in stages. May geologists who disagree please give real supporting evidence for disagreement. Logistics must always supersede speculation. In my theory of hydrostatic-compression from the west caused the Western Overthrust Belt, the San Andreas fault and all major faults of the earth, including those in the Overthrust Belt, are all right where they should be.

There is both shock and amazement as we learn more about the great Western Overthrust from Alaska to Mexico, possibly Cape Horn. Yet, in annular geology, everything is just as it should be. It would seem reasonable the Overthrust was formed during the jurassic to tertiary times. Many geologists and sedimentation experts agree a great compression force from the west drove the overthrust east over the continent causing massive slabs and faulting to form. The question the old-school geology does not answer, is what caused the great compression force? It seems only logical this force was supplied by a great hydrostatic compression force caused by ocean augmentation from a ring fall of water around the earth. "Isostasy" must be a true concept because of the direction gravity is acting, we get an overwhelming lateral force from the addition of vast amounts of water to the oceans from a ring system once around the earth. If we are to accept old-school geology, no one would drill for oil in the Western Overthrust Belt.

Some have referred to the Overthrust as a myth, a bottomless pit. This would be true if we try to account for hydrocarbons being present in it from the vegetarian geology concept. It is next to impossible to define where the basement is in each given area which is as it should be in annular geological theory. The Overthrust is a very puzzling geological formation. The accepted geological explanations are 30-50 years old, and will be corrected and updated as new drilling disproves old theories based on guess work only. Some say the structure is a nightmare. Why should it be a hydrocarbon province? I will say there will be both disappointment and exaltation in oil exploration there. Old-school geology does not answer many questions about the Belt area. It is a totally conglomerate geology, where standard seismic readings will not apply. A new interpretation of seismic understanding may have to be developed and even then they may not reveal a dependable science. Isn't it possible, some oil beds may be upended and difficult to find, not only with seismic as well as drilling? Conformity will be the exception rather than the rule. Geologists and geophysicists will change their minds many times.

One Belt drilling project in Idaho is looking for Precambrian rocks at about 19,000 feet deep. If we apply standard geological thinking to the Overthrust isn't it a very bleak area to look for hydrocarbons? Then on the other hand, known oil seeps have been found in many places. If Belt rocks are supposed to be over 30,000 feet deep in an area, we may wonder why Mississippian formation may be found at about 1,700-1,800 feet deep. Why should we see Paleozoic outcrops in an area which should be 30,000 feet deep Belt rocks? It would seem possible as the Western Overthrust was developed a vast amount of rock and other material was forced lateral under the continent from the ocean on the west. Do you have a better explanation? Why should there be Paleozoic outcrops and others in a belt where such should not be the case, unless we accept annular geology as the developer of this upheaved formation? Carbons were deposited to the earth from an earthly ring system, as were other materials, over millions of years, so oil is still where we find it. We cannot say this Belt is metamorphic any more than we can say it is Precambrian. Metamorphism is the great fallacy of geology. It must take place before or during rock development and placement, never after, as does elemental segregation. If Belt rocks are neither a source or a reservoir for oil and gas, how did the oil get there? If an earthwide age such as Cambrian got elevated and upheaved, when and how is the question?

May I suggest, and I suggest only, that the possibility does exist, that the Great so-called Western Overthrust is not an overthrust, but rather an upheaval from earth material being forced laterally under the continent from the west. Just a thought.

If this very complex geology created folds and faults favorable for oil (hydrocarbon) traps, at what stage of development did vegetarian geology place the hydrocarbons in the beds and traps? If older rocks in the Overthrust overlay younger rocks, will oil and gas be found in the younger rocks? If hydrocarbon existed in the younger beds was it placed there before the older beds overthrusted the younger? If not, how did it get down under there? Yet to be answered, did the oil beds in the overthrust part of the formation exist in them before the Belt was overthrusted? It is evident carbon and oil had to be in place in them before, or after, we only have these two choices. Where did the fish come from to create the oil fields during such an age of upheaval? This formation was not stirred together like a cake. It has to be a fact, most of the hydrocarbons in the Western Overthrust Belt were in the Belt before it was overthrusted. This will account for distorted and upheaved oil beds in places least thought to be found.

Further, if oil and gas are found in formations below the overthrust material it must be true, these hydrocarbons were placed in that bed before the overthrust took place. There is a logical procession in annular geology which cannot be reversed. If the great Western Overthrust were a pancake, no matter how many times we turn it over, we would come up with the same answers. There seems to be an accepted premise in both geology and paleontology that any action, or fossil find, must be the same age as the action of the rocks or the fossil found. Is this necessarily true? Give this some serious thought. Some experts place the Overthrust as occurring 100 to 150 million years ago. I will not tie myself to a time estimate for the Overthrust action. Does such an action or age of rocks need to coincide? No it does not! How could a rockbed be forming at or even near such an action as the Overthrust? It would seem in this case the rockbeds must predate the action by millions of years or the overthrust action would have nothing to work with. Both jurassic and tertiary ages had great earthwide upheavals. The age of the beds had nothing to do with the time the action took place. I will say this, that the Western Overthrust may very well have occurred in the more recent geological times. Doesn't the fossil record on them support this? Why was the earth once tropical and semi-tropical beyond the Arctic Circle? This was just pre-deluvian as the climate changed suddenly.

The fossils now found on the Western Overthrust did not necessarily need to develop after the thrust action took place. They could have lived and died during the thrust action and just took a free ride on the top where they are now found. Maybe some time deep fossils may come to surface in drilling if they did not get ground beyond recognition in the thrust action. Such a geological conglomeration should produce some fossil conglomeration too, which may answer some troublesome questions.

The answer to much of our riddle is found in annular geology and hydrostatic compression due to world and ocean augmentation of water and material which kept the continents and mountains heaving higher until the last great upheaval caused by the last downfall of water to overwhelm the earth. Western Overthrust geology presents a situation wherein vegetarian geology just cannot survive. How much of the Cambrian beds were overthrusted is the question? Was all of the Precambrian overthrusted in the overthrust? If the overthrust left none of the oil bearing rocks to the west, will oil ever be found far to the west of the Belt? There is oil drilling off the coast of California with good oil finds. Could this area then have been overthrusted? If so, why didn't the thrust carry the oil beds with it? Will oil be found off the coast of Washington state, or in western Washington? Could the area of Banff, Canada, be overthrust, or just upheaval, or some of each? The stratigraphic column of this area does not have conformity to original layment, nor should it have in annular theory. These are good questions; perhaps the issue of Belt sedimentation has been decided too hastily and should be relegated to mythology. An entire new format must be developed. Who then, is willing to admit error?

These great conglomerate beds of geology bear unimpeachable testimony of cataclysmic birth. If glaciated, when and from what source was the glacier? The puzzle can only be solved in annular geology. Water, oil, coal and ice were delivered to the earth from its one time great ring system, such as Saturn still has. The fiery and igneous earth distilled and supplied all of the carbon in the form of coal and other forms we now find, long before the slow decay process of vegetation ever took place.

Fire held control. We know hydrogen and carbon were two all-abounding elements in that primitive furnace. It is known that hydrogen and carbon, thus conditioned, actively seek combination, and unless they passed through a sea of free oxygen on their way to the skies, they arose as oily products. Lack of free oxygen at that time would not allow oxidation. It is impossible to avoid the conclusion that fuels were thus formed and went to the skies as unconsumed carbon, later to return to the earth in many stages and kinds of oil, coal, and carbon. We are now mining this coal and drilling this oil. My reference to this great complex geological area is mainly western Montana and Wyoming, and surrounding states and Canada. The carbon development process of annular geology I apply to the entire earth. I have seen basement depth figures for the overthrust areas of 30,000, 60,000, even 65,000 feet deep. It is my claim there is a noncontiguous oil, carbon, and mineral ring or belts all of the way around the earth, even under the oceans. Because oil was not placed in the Overthrust Belt by vegetarian geology, there is no reason why vast amounts of hydrocarbons could not be found in the Western Overthrust Belt. In this case, we must never forget that the bit still proves the presence of oil. Just as a building is built from the foundation up, so was the earth stratified.

ELEMENTAL SEGREGATION & ASSORTMENT

The Crollian theory supported by Geikie and restated by Dr. Gould as truth must be rejected as out of harmony with sequence of known physical laws of nature. (Geikie's *"The Great Ice Age."*)

It seems factual that the carboniferous age must have evolved and developed simultaneous with many of the major geologic ages. Silurian lime was laid down. Stupendous devonian rock beds were placed on them. Permian oceans left its load. A triassic flood deposited its load. Then the jurassic made its appearance, followed by cretaceous. As Vail says: *"If the cretaceous deposits were only local, we might attribute the change to local causes. But it was a change that left its way-mark the circuit of the globe."*

It has been a constant puzzlement to organic accepted old-school geology with the known build-up of the geological strata of the earth. Why then, is there such a vast elemental difference from the cambrian up through to the cretaceous, if we have a single source of material and water, in a single environment, with a single process, erosion build-up? If this be true, what was our force and machinery for elemental separation and segregation, in such stupendous geological beds? We must all admit if the oceans are to build lime stratum, which they did in the geological ages, there must be a source of the lime. Where will we find such a source of lime to supply this measureless fund, independent of both carbonate and magnesium? All geologists know the waters of the cretaceous period were radically different from those that preceded it. In some parts of the world cretaceous chalk beds were deposited. Then calcareous beds radically different from any other lime-formation. From what continents did the ocean get its lime that it should be so different from every other limerock? Where was segregation done? Who or what enforced it? Unless lime is contained in the ocean as a solution the stratum cannot be formed. We must have the blocks to build with.

The silurian age is a world wide age. Why should there be a silurian age with such radical change in geology and ocean waters? Geologists know that the change from eozoic (Precambrian) to silurian was a worldwide one, and how could it be possible for the same waters that gave origin to the Potsdam sandstone at the base of the silurian system, supply the stupendous world casement of lime unless there had been a vast augmentation of lime-waters? We still must consider the oil content of some silurian rock. An oil well in North Dakota drilled into the silurian at 14,000 feet deep made over 1,000 barrels of oil and 1,700,000 cubic feet of gas per day, on a 20/64 inch choke. It makes no difference how the geologist supports his claim that the lime beds of the silurian were derived from older continents, he is continually arrested by the demands of law. We still must answer the question of elemental segregation. Look at the vastness of this formation. We cannot relegate it to the playground of "uniformitarianism geology" accepted by learned men. If this be the case, that is exactly what it would be, "uniform." The opposite is true!

To me this is the most important chain of thought and much more could be written. Perhaps I am boring you though. Let's get to the point! In our world-wide continental strata-building process, what supplied our segregated elemental material to the surface of the earth in such stupendous amounts? It was the annular rings once around the earth. Because the geological ages were built from the bottom up to surface, the segregating of the material had to be done before it was placed. Elemental segregation had already been done in this ring system. The varied materials for this great strata build-up was supplied by the aqueous vapor rings from space which already contained them. Look at Saturn, there is segregation in its rings. The great primitive aqueous envelope surrounding the earth would naturally form a world-wide bed of matter. The material needed was thrown into the ring system at the beginning, in the igneous earth. See the pictures in this book.

Just as the potter turns his table, he can only make circular objects, yet he may slowly work in different material out to the surface. Compare this system to a turning wheel. We could pour some water on it and some would stay on all around the wheel. Next pour on some oil, again some would stay on all around it. Next put on some flour, some sand, some of many things. If the wheel were not turning, we would not have the same effect. The segregation was done by choice from the outside. So it is with the strata build-up of the earth. Elemental segregation was done by heat and centrifugal force by a rotating earth. Elements were systematically separated in the ring system around the earth and today we find them right where they were placed in this orderly function. A world had to be built some way, didn't it? In a lifetime of research, I have not found any idea which can successfully replace this one. Perhaps you have never heard of it.

Therefore we are led to the conclusion there is a carbon ring all the way around the earth. It is this far-reaching and sometimes universal change that directs us to the annular-cosmic matter of primitive time and ages. Vail also said: *"If men of this age refuse to use this key, other men will gladly embrace the opportunity when we are in our graves."* Vail died in 1912 so the time is now. Perhaps in the snowbound polar deep, packed in eternal ice, ice above and ice below, we will bury the old theory and erect the new. The annular theory demands it.

On my desk I have two pieces of rock. One is oil shale from the western U.S., the other is a core sample from 8,000 feet deep from the Madison formation in the Williston Basin. The core sample is very hard rock with tight porosity, perhaps anhydritic shale or limestone. The oil shale I have is very hard and tight too, it is from Wyoming. If I were to show you these two rocks and ask you where they came from and how much oil they had in them, you may just laugh and say absolutely no oil and the rocks both came from the same area. The rocks do appear almost identical, and they both contain oil. The truth is there are billions even trillions of barrels of oil locked in oil shale deposits around the world. Likewise there is just as much oil in tight shale, limestone, and other oil bearing rock deep all over the earth. I

am making a point and it is not that we no longer have an energy problem because of these vast untapped oil deposits. This oil may never
be recoverable and if so the cost will be very high.

My point is that these vast oil deposits do exist all over the earth
and under the oceans as yet undiscovered, add to this all tar sands of
the earth, also all recoverable oil past and present, from shale in
mountains to 30,000 feet deep, when did the marine life live that the
vegetarian geologists claim made all this oil and hydrocarbons? We
are talking about trillions of trillions of barrels (42g) and trillions
upon trillions of tons of gas, oil, and carbon all of it different in some
respects, distributed all through the earth's crust, perhaps to 50,000
feet deep when proven. How did fish imbed all this oil in solid rock? If
fish placed oil in this rock hundreds to thousands of feet thick, when
was it delivered to the rock and how? One piece of oil shale rock I
have is stratified every one sixteenth to one quarter of an inch thick. If
we say absorbtion, then why did it seek the hardest rock for a home
when ample softer rock was available? Would this be a reversal and
violation of physical law?

Scientists have boxed themselves into a corner. We have an even
greater hurdle to cross in this theory. Isn't it a known fact that nothing
organic will turn to oil or hydrocarbon in the presence of free oxygen?
In order for our agent fish, and other life to live so prolificly so as to
produce all of this overwhelming amount of oil, there had to be oxygen present. Yet, with oxygen present no carbon can form! It would
self-destruct in its own process, oxygen would devour it in spontaneous combustion. Smoke from a chimney dissipating in the atmosphere is secondary combustion in the presence of oxygen. How
can they cross these great unheard of boundaries without accountability to law? In the hundreds of millions of years the incalculable oil
and carbon beds of the world were developing, from shale mountains
on surface, down to 30,000 feet deep, did we first have a million years
of oxygen, and then a million without so carbon could form? How
could such a mythical illogical process exit? It staggers the mind to
lunacy when we try to apply the vegetarian theory to the real world
around us. The 30,000 foot gap we are concerned with over millions of
years is a large gap for the puny agent, some fish to fill. We will soon
see oil drilling in 8-10 thousand feet of water. I feel they will find oil.
When did fish place it there?

Will science in an attempt to cross this boundary and get out of
the corner they are in, now apply the quantum jump to this geological
fiasco, as they have done in genetic evolutionary speciation?
Everything must have a counterpart. With no reflection on you, my
intelligent reader, isn't it true, intelligence must be cultivated. Perhaps

then, ignorance must also be cultivated. With these two great reaction forces working against each other shouldn't the truth eventually come out?

THE MAMMOTH

The Smithsonian Magazine is a very interesting and educational magazine. I will refer you to a very interesting article in the December 1977 Smithsonian, about the mammoths from Siberia and the Ice Ages. The Siberian scientists give seven possible explanations as to how these animals died and are frozen in the earth, none of which are plausible. Each of the seven take time and the physical facts are, these animals died suddenly and in large numbers, over a very wide area of the earth. Sometimes hundreds of different kinds of animals were found packed together in one place. One tusk could weigh more than 185 pounds. Siberian scientists feel only a small amount of these animals have been found with great numbers still buried or washed away. Most were mammoths, but occasionally other animals were found with them: Arctic hare, wolf and wolverine, horse, bison, reindeer, wooly rhinoceros and cave lion. The big mystery is, why did the mammoth become extinct along with the cave lion and the wooly rhinoceros? My explanation for this geologic mystery is the great down-rush of water, when the last annular ring on the earth fell into the earth, with its water and ice and snow at the poles. Siberian scientists went on to say, according to this interesting article, that landslides and erosion caused this death and destruction, and perhaps some animals fell through ice. None of these explanations can be right because of the time element, and there was no ice there at that time to fall through. It was semi-tropical.

THERMODYNAMICS

At this juncture in eternity, in the vast empire of universal dynamics, is anything ever diminutive or expandable, or is it just transitional, within the absolute bounds of displacement? Perhaps we are locked to a permanent fixed amount of mass and energy, which only trades back and forth within itself, whereby the universe may sustain itself much as it is for the rest of eternity.

In thermodynamics do we not see a reversible transformation of heat and energy by laws governing such conversion of energy? Consider the remarkable transformation in plants using sun as energy in the chemical process called photosynthesis. One book I have on the subject of energy says: *"While catalysts may seem to go against the grain of nature, they are careful to obey the Laws of Ther-*

modynamics. When hydrogen combines with nitrogen in the presence of iron oxide, for example, the resulting ammonia is a more stable molecule. Without the application of external energy, no catalyst ever induces a reaction to go 'uphill.' The inescapable First Law of Thermodynamics tells us that the energy required to break up a molecule is equal to the energy obtained when the molecule is put back together. The remarkable thing about photosynthesis is that the plant achieves something a chemist finds hard to accomplish with 5,400 degrees of heat; it splits a water molecule." Perhaps we should now add the green pigment, chlorophyll, at this time, also absorbed from the sun, and we have a process hard to duplicate or beat.

Doesn't this support the theory of universal permanency and stability? Much scientific information has been gleaned in the last 80 years; however, isn't most of it only an embellishment of already discovered and known fact! Isn't this borne out by Einstein's theory of the equivalence of mass and energy. According to Einstein's equation the conversion of energy-to-mass was absolute. This was later proven to be correct though Einstein did not work out or prove his equation inside a laboratory, which astonished his peers. Einstein said: *"Physics is a logical system of thought in a state of evolution. Its basis cannot be obtained merely by experiment and experience. Its progress depends on free invention . . . I haven't the faintest doubt that I am right."* His equation holds true in the transformation, whether from mass to energy, or energy to mass. When asked how he proposed to release all this hidden energy, Einstein said: *"There isn't any indication that the energy will ever be obtainable."* Later it was proven it would be though, and most of it not to the blessing of mankind.

It does seem we are governed by absolute laws. Einstein also said, *"Nothing could go faster than the speed of light."* Another absolute (186,272 miles per second). According to Einstein's work, (1905) there is nothing in the universe that we can be sure is stationary. May I add, why should it be, isn't motion and attraction the force holding all things together?

About the time Einstein was doing this work, another scientist, Isaac N. Vail, originator of the annular theory of planets, described the Planet Saturn, just as found to be by the Voyager Mission. Vail did not have the benefit of the laboratory or the findings of modern space missions, yet by his understanding of physical law, he brought the world such advanced knowledge as did Einstein, it was not accepted. Isn't this the way it has always been? The theory that all chemical fuels, coal, oil, are a by-product of the carcasses of decayed bodies of plant and animals must be discarded as wrong and not provable. How could this process place minable coal beds as deep as a mile or more into the earth and sometimes interspace them with oil and/or gas is a question no one has ever answered. Also placing other kinds of mineral seams and stringers not related to coal above or even below the coal beds. How was the sorting done, please answer me? Centrifugal force could do this mighty task but not slow creeping sedimentation and erosion. Sedimentation did play an important role in world building geologic process when applied to annular geology. It was rapid at times. Tree fossils buried in coal seams is a good example.

CHAPTER 10

ARCHAEBACTERIA

At present, I am giving some thought and study to genetic improvement by chance mutations. Because enormous numbers of chance mutations would be required for even a small improvement it seems to me it would move backward, not improve or evolve up. It is loaded with missing links. Also "Archaebacteria" as a research of bacterial microfossils, is full of missing links. It seems to be an entire missing chain to me. Bacterial microfossils are not very informative structures it would appear. The evolutionary status of archaebacteria is an untested conjecture. Do they really have a form of metabolism, or is it just a chemical arrangement; with no known line of demarcation. Good questions, we need some answers, which may very well destroy the theory.

Dr. Carl Sagan from Cornell University, and Dr. Stephen Gould appeared on the Donahue Show, May 27, 1981. These two men are very well informed on their subjects. On this show it was said, the universe is 15 billion years old and the earth is 4.5 billion years old. No argument here as no one can prove it and what difference would a few million years make, anyway. It was claimed the stuff of life is very simple and came together earlier than once thought, by accident. Both Sagan and Gould say absolutely, that biological reproduction evolved by chance over millions of years time. This is not a provable or testable claim. It is just plain wrong. We may ask, were there millions of years of partial pregnancy then? If so, how did life get to where it is today? The makeup of our body has some 100,000,000,000,000 tiny cells, each one very intricate and well designed. Then, with genes and DNA, they determine everything about us at conception, down to the color of our hair. This all had to be programmed like a computer has to be programmed. If the RNA and DNA are programmed to perform as programmed molecules must do, in designated patterns, who did the programming? How did creatures see before the eye was fully developed, or function before the brain was fully developed? As anyone knows, any damage to the eye now will cause blindness, or to the brain causes death. How did this process survive a hostile environment over millions of years?

Science has left far too many questions unanswered, or covered over with unprovable theories. I enjoy reading Scientific American Magazine, it is very interesting. April 1981 issue has an interesting article on the Origin of Genetic Information. The first part of this article is so loaded with disclaimers, why read it at all? Many wrong remarks are made as well. One is: *"Acids and proteins that have survived three or four billion years of molecular evolution."* As science claims the earth is 4.5 billion years old and that far back the earth was so hot no living thing could survive. Next the article says: *"It is impossible to re-create the actual stages of genetic improvement because enormous numbers of chance mutations were tested and discarded during early evolution."* What has happened to our marvelous programmer who does such a superb job now and was only guessing and blundering four billion years ago? This theory just cannot be correct. Next an excuse is offered: *"It was therefore necessary for the first organizing principle to be highly selective from the start. It had to tolerate an enormous overburden of small molecules that were biologically 'wrong' but chemically possible."* Can you, in your logical mind, support such a claim?

Next this article attempts to set the stage for the origin of life by transition from inorganic to organic material. Especially those who teach vegetarian geology try to say the earth had the same temperature billions of years ago, or where would they get the organic life which they say made all of the coal and oil, etc., of the earth. The article now says: *"The stage was somewhere on the primitive earth, which had a temperature not much different from what it is today."* This is not a correct or provable claim. It is a known fact the earth was very hot, even igneous, that far back, and would not support any kind of life. Another wrong claim is made. *"The total amount of potential organic material was immense."* This is not true as it was too hot for organic material or life to develop. Organic material did not make coal. Next the claim is made that, *"Fermentation would have been adequate until the advent of photosynthesis provided a continuing energy source."* They admit their dilemma by saying there had to be some kind of conversion of solar energy to chemical energy to fill this long gap. This one really throws me. I did not know there could be fermentation before photosynthesis and organic material and free oxygen. Did you? For life to evolve by brewing in this sea of soup, as they call it, just does not seem reasonable.

There seems to be a desperation to push the origin of life farther and farther back in the time calendar — it has become competitive rather than factual. We see this in a recent fossil find they have named Lucy, apparently the remains of a creature. Fantastic age has been given this fossil or whatever. I have read articles in scientific magazines and have not been able to make head or tail of it. The claims are just ludicrous. It is said "Lucy" is the oldest complete preserved skeleton of any erect-walking human ancestor ever found. It is called the beginnings of humankind. Lucy is approximately 3.5 million years old they say. The fossil was found at Hadar, in Ethiopia. Lucy was only 3½ feet tall although fully grown they say. They decided this by her teeth. After reading many articles about this fossil find in scientific magazines and leading news magazines, I am not convinced or impressed with the claims made as they are not provable. Just someone's opinion. The age of the fossil it seems, is based on a tricky and doubtful method of lava and volcanic-ash dating process. If the ash is 3.5 million years old, this would not prove the fossil is that old. The fossil was found right on the surface with some bones exposed. If we are to believe the figures of erosion geologists, bones would not lie there in the open for 3.5 million years. They would be destroyed, washed away, or be buried far underground. From all of my reading, all that was proven is a fossil find.

Some now claim microorganisms have been in rocks 3.5 billion years old in Australia. If some form of microfossil has been found in old rocks, who is to say the fossil is as old as the rock? Is there some confusion in the line of demarcation, if there is one, between microfossils, bacteria, and chemistry. In geology we sometimes find structures in reverse order to normal. The June 1981 Scientific American on "Archeabacteria" says, *"Bacterial microfossils are not very informative structures; little can be inferred from the prints of a small sphere or rod."* An interesting scale is given which proves very little. As one expert says about the western overthrust belt, we see great plates of rock, some of them four miles thick, are thrust up over each other like toppled dominoes. Cataclysmic things have happened to the earth's crust in more recent geological times, as is easily proven. Drilling cuttings and core samples from this area coming to surface would have once been the surface, at four miles deep. What provided the energy for this fantastic overthrust of a large part of the United States and many places all over the earth? It was provided by the igneous earth when matter was thrown into a great annular system of the earth over millions of years. A large part of this energy has now been spent, or transferred I should say, and nonrenewable. The four mile estimate for the thickness of the western overthrust belt may be too much. Time will tell if this is so. All things are of necessity sequential, and become consequential. Perhaps we have some missing links in Archaebacteria too. As Einstein said, *"Order, not chaos, rules the universe."*

CHANCE MUTATIONS

It does not disturb me that many scientists teach the theory that favorable chance mutations developed all forms of life. It is disturbing the free blanket coverage these scientists get from the media, to propagate their unprovable claims. Their evolutionary status is just untestable conjecture. I am thinking of an example to show how impossible genetic improvement could come about by enormous chance mutations in an environment hostile to its success. We may compare this theory to someone trying to climb up a sheer straight up canyon wall, with 100 bad guys waiting at the top to step on his fingers each time he got to the top, sending him falling to bottom each time. Then we must consider how many times our climber fell back part way or to zero on the way up before he made it to the top. The 100 bad guys, like the hostile environment, would not allow him to get over the top. How many failures would it take for one success? Yet in this simple illustration we need billions and trillions of successes to accomplish one single cell, not to say an entire world of life. Are the promoters of the chance genetic improvement theory by blind accident, now going to claim, once this action found a start, the chain was never broken! If so, this would be counter to, and contrary to their own position. If there are any breaks in this hypothetical evolutionary chain, how many are possible and how far back in the chain must we go for a new start? In many, the break could be at zero, so wouldn't the new start be zero? Would you like to try to supply some of the answers? I have many more questions, but if we can find the answer to these that will be a good start. What really was the source of food for small fossil creatures? Was it absorption or ingestion? Was it metabolism? From what?

In the precarious assumption of genetic improvement by chance, who could prove which direction it is moving, up or down? Could it not move backward just as readily as forward, even more so, perhaps to extinction? For a basis of a consistency factor, in Einstein's equation of the equivalence of mass and energy the conversion of energy-to-mass was absolute, and held true whether from mass to energy, or energy to mass. If an outside force was being applied to genetic manipulation over billions of years, where did it come from? There are too many omissions.

ROCK CYCLE THEORY

Going to another subject, is anything ever metamorphic without an outside force. Consider the rock cycle theory, "New rocks from old." Does anyone really know which direction the process is moving if it is true at all? Do the laws of thermodynamics ever violate physical law, without an outside force? Why do we have so many scientists trying to sidetrack the actions of true physical law, and have they really succeeded? No! They have not, they rely on assumptions and speculation, without proof.

In the rock cycle theory, it is assumed, rocks are forever changing back and forth through the rock spectrum. In this unprovable and doubtful metamorphic action, who would know which direction it is moving? Where does it start and where does it end, before reversal in the cycle of change? How far downward does it travel, perhaps to the center of the earth? If not, what decides when it should reverse itself? Will it travel through an oil reservoir and leave it intact and not release the pressure, destroying the oil bed for future production? Does it move through coal beds changing the coal to something else and then on the way back change the coal back to coal again? True, we may find some rock blending where rock beds are thrown together such as a schist. This blending must have happened when the beds were developed and placed where they are, yet this is not proof they have been changing from one to the other. Who will decide which direction the change is moving may I ask? Would we argue when a boy falls off his bicycle in the gravel street and there is gravel and flesh enmeshed together would we say the boy will become all gravel eventually or the gravel change to flesh? The "rock cycle theory" is another one which must be discarded as against reason and logic. I have seen on the other hand, oil well core samples, where the rock on top of the oil bearing rock is instantaneous with no blending at all. This is common. Could this perfect line of demarcation in rock survive one rock cycle invasion, or hundreds as is claimed? In all frank honesty; isn't the theory almost laughable? You decide, please! (See page 244)

Grain alcohol and oils from sunflowers or corn, as well as other organic substance may be used in cars and tractors as a fuel for power. This is solar energy, and does not prove oil and carbons of the earth are made of vegetation. All material substance is energy, all we need is a conversion process to a useful need.

When Einstein was very young, he applied for admission to the Swiss Federal Polyechnic Institute in Zurich. He was turned down because of inadequate training in languages and other fields. Later, however, he was admitted. It was with interest to note on the Cosmos Show, it was revealed, many scientific discoveries were made by men with little education or background for this. One was a banker's son who did not want to go to school. He drove mules to haul material up the mountain to build one of the great observatories. After it was finished, he stayed on as a janitor, and later learned to operate the telescope. Later it was he who made some of the most important cosmic finds of all time. The planet Uranus was discovered in 1781 by an obscure amateur astronomer and musician by the name of William Herschel. It too, is a ringed planet.

After Dr. Albert Einstein developed the equation for the equivalence of mass and energy, his peers doubted and questioned him, as he had not tested his theory in a laboratory. His answer was, *"I haven't the faintest doubt that I am right."* I do not mean to compare myself to Einstein, however, I feel just as he did upon his discovery, I do not have the faintest doubt I am right in my theme on vegetarian geology. It will be proven to be correct.

Perhaps one big problem which has persisted for centuries is the attempt to blend partial truth, conjecture, and erroneous theories, with fact and truth. The erroneous ideas and teaching of Aristotle set science back for about 2,000 years.

In a book I have it is claimed that gigantic trees grew 250 million years ago in shallow swamps. As they died and sank to the bottom, they became the homogenous black mass of a coal seam. Why are some coal seams over a mile deep and others right at surface and some only a quarter inch thick? Perhaps it is true, in the chemist's retort, they may be able to make very small amounts of oily substance from certain plants and organic material, or even a swamp by removing the oxygen. This puny process did not make all of the oil and coal beds of the earth miles deep.

Perhaps another consideration of carbon should not be overlooked. It is claimed that diamond is a form of carbon. Diamond is very hard if not the hardest substance known. Would you care to say this material was organically placed? Diamonds as you know, may be found right on the surface of the earth and they may be mined deep in the earth. Would you say their solid hardness is caused only by pressure? What made the surface diamond hard? A news item now says scientists in examining meteorites and interstellar dust discovered they have carbon-rich compounds. Was this carbon organically placed? Meteorites seem to be carrying another message.

UNIVERSAL EMBRYONIC MATTER

Will ultimate answers be found in physics, mathematics or philosophy? Are we governed only by laws of chance? Is a universe with humans in it, inhuman? Why is psychiatry a booming business in a "brain age?" Should not science and reason provide the ultimate answers for life? Are we to assume that replica creatures kept coming into existence without due process of biological reproduction for billions of billions of years? Today scientists, evolutionists, and religionists, deny even the virgin birth of Christ. Yet the Bible is only asking we accept one time, and that with the power of the almighty God. On the other hand, science is asking us to believe billions of unfathered virgin birth creatures came into existence over billions of billions of years until biological reproduction by chance established itself so forceably, no one would dare claim life may be reproduced by any other process. Not only that, science claims this was preceded by billions of years of chance mutations of genes from a single cell until the higher forms of life by chance came into existence, and for some unknown reason, abandon chance mutations and virgin birth, in this senseless chain of events, and adopted biological reproduction forever more.

In the beginning there was this "big bang" as they say. Thousands of volumes have been written covering what happened from then to now. Endless words and theories have been developed on just the few seconds after the "big bang" and what happened then. One reader of Britain's *"New Scientist"* magazine had the courage to ask: *"What happened seconds before the 'big bang'?"* I am sure when the sober calculations of man are brought to bear upon this great question, it must gravitate toward an answer to our British reader's question: *"Why should sufficient embryo matter to make the whole Universe suddenly appear out of less than nothing? Why should this matter obey fundamental laws which have made all subsequent development of the Universe possible? Answers please, or just hypothesis."* Which is more in accord with the laws of universal evolution? Doesn't this bring us back to an old standard question: Who created, who created God?

OCEAN AUGMENTATION

I am a very privileged person, in that I have known for over thirty years about annular theory. That the rings on Saturn would be many, even hundreds, yes, thousands, and would be filled with material and substance from the outer ring down to the last ring at the base of the planet. Also, some rings would be braided and eccentric, and would appear to violate orbital laws. Too, that there would be many moons on Saturn causing a tugging effect on its rings. And that there would be elemental segregation in its rings, just as there was in the rings once existing around the earth. I have talked about this to my friends and acquaintances for over 30 years.

The most outstanding scientist, Isaac N. Vail, had said these things and taught them more than 80 years ago, yes, as far back as 1874, and even before that. He wrote many books and papers, and gave many lectures proving his theory. He described exactly how Saturn would be, and now the Voyager Space Missions have proven his ideas and concepts. The space missions and flights have been one of the greatest accomplishments of man. Knowledge of the Universe has been greatly increased.

We find signs of glaciation in all the geologic ages, except maybe the very oldest. Since no competent source of snows can be found on earth, Professor Dawson declared that a great continental glacier is a physical impossibility. Yet we do know there are glaciers on the polar continents, and there is evidence of glaciation in the geological ages. A great continental glacier, without a competent source! The mistake of the glacalists is that they claim the earth now possesses an adequate source of snows when every feature of philosophic law is against such a claim. We must again turn to the fall of annular rings which contained the needed material. How else may we account for the universal extermination of life-forms, whereby whole races of animals on different continents vanished. This had to be catastrophic, universal. We cannot satisfactorily account for universal uplift of strata except by augmentation of volume and weight of water and other material. The fund of annular matter was not yet exhausted.

There was great permian upheaval. Great disturbance is evident in carboniferous and triassic. Some of the more recent periods were characterized by the most stupendous revulsions and upheaval. The Alps, the Andes and Himalayas were lifted in more modern times. The jurassic had one of the grandest mountain-making upheavals of the world. One expert says the coal period was one of unceasing change, with alternating destruction of all vegetation and life. Why were such wide reaches of coal beds formed in the lap of the glaciated world, if they say vegetation built them?

Experts say the Ural Mountains as well as Australian upheaval was permian also. These oscillations show contemporaneous movements on both sides of the Atlantic Ocean. There is evidence of contemporaneous displacements in all continents. We must have a world-wide force to do this. What happened to crocodilian life forms so widely once on earth? They were caught in an annular downfall, it is evident. The final result, a new distribution and condition of oceanic waters, refrigeration and consequent extermination of species.

We must think seriously and deeply to get a picture of all of this in our mind's eye. To lift mountains and continents is no small task, yet we know it did happen, we are here to view it. We know the forces that have upheaved the continents have always acted from the oceans, and that they have acted again and again, after it is once applied, except through force stored up as potential, either by additional waters or solid matter. My thought was, the oceans cannot use a force twice unless new energy is supplied. The pressure upon the ocean-bottoms must have become so great as to force strata laterally under the continents. We see that many times the upward movement of the crust has been simultaneous in differnt continents. We absolutely must have forces employed which are competent, and causes adequate to accomplish these grand results, we do see them acting in harmony with law; in harmony with nature in every way, astronomical or geological. We do not need to wonder longer at the constant oscillation of sea and land throughout geologic ages. The oceans of the world now maintain continental stability. Continents cannot drift but very little. We no longer have annular rings on the earth to cause great change. Ring pictures here shown were developed by Isaac N. Vail over 80 years ago. The pictures are very helpful in understanding Vail's proven theory of annular planetary development. It stands solidly today as when it was first developed. Space missions will prove it correct. (Pictures in back.)

Any geologist who attempts to define the earth's crust and topography without admitting a great flood of water and ocean augmentation will become entangled in a maze of unexplainable inconsistency. There is both astonishment and deep satisfaction revealed by the hoary rock-volumes which strata of the earth have been crumpled and upheaved. Geologists of the world know that this was a continent-making process with great crust-disturbance, extending back to archean times. Increased pressure upon ocean floors over millions of years of annular canopies falling into the earth with strata buildup caused upturned beds resulting in disturbance, planting one strata upon the ruins of a former world. The materials had to be supplied from somewhere. We cannot make bricks without clay, or mortar without mud. Each geologic age had its own characteristic matter and elemental content quite different from the earlier one. Segregation is not possible after placement. This has been a stumbling block to science for years, just as the presence of free oxygen has been to the organic formation of carbon, and science. These facts must be dealt with. Science in the future will prove the planets of the solar system contain vast unheard of amounts of primitive carbon in their makeup, not of vegetive derivation. I have explained this to people for over 30 years.

It is self evident that periodic glaciation of the earth over millions of years time had to be from annular water and snows, it could not be a terrestrial source. There would then be continuing and extensive land upheavals, until the last downfall or Great Flood. Each of the geological ages are so different from the former. Each has its own characteristics. Are we to argue and claim that each succeeding age borrowed its building material from its predecessor? Consider an older age the cambrian, did the silurian gets its material to build itself from the cambrian? Did the devonian borrow material from silurian, did permian get material to build from its predecessors, did triassic and jurassic considering it is world wide? They are all different from each other! How can we borrow from someone who has nothing to lend? The waters of each age are also different. It must be staggering to science since they have had things in reverse order for so long. If the waters of the ocean did not contain oxygen, where would we find the whale? Fossils, when they do exist differ from age to age and stratum. How could this be if one borrows from the other? Humor sometimes helps prove a point — I found this one somewhere: My poultry-house contains a chicken thief with a "large percentage" of fowls in his possession; this explains how my chickens got into the house? Would we conclude the feathers left behind were made by the thief also? Yes, it is these far-reaching and sometimes universal changes that direct us unerringly to the tellurlo-cosmic matter of primitive times.

DISTRIBUTION OF EARTH'S MINERALS

The subject of variation within a geological age, and transition from one age to another is a must in this theme. There are many geologists who know these details very well, especially those who work on oil drilling rigs and study bit cuttings coming to surface. We may wonder why the passing from one formation to the other is quite abrupt, such as from silurian to devonian. (See Fig. 15 on page 244.) Other places and ages it may not be abrupt, but rather gradual. In the annular theory we see a gradual build-up of a geological age at times and again more abrupt as the great earthly ring systems deposited their loads to each age. Geologists have found and named many identifiable geological structures within a given geological age. Deep oil drilling has added much information to this work. This is evident in reading oil well drilling reports from thousands of wells. Affecting the great geological beds and transformation from one to the next would be continental upheaval from annular waters and snows being deposited to the earth. Glaciation would originate from the snows and water of these rings and not from ice caps moving down from the poles. There were epochs of change and disturbance which is seen in moraine deposits of some ages. In passing down through the ages, we find evidence of unconformability we should not be perplexed. In annular theory, there was a source of glaciation competent to transport materials. This will account for conglomerated beds intercalated between coal seams and many other hard to understand puzzles.

It is manifest to even a common observer that the strata of many ages (carboniferous, too) were bent, crushed and folded in many corrugations even after coal beds formed. I am always harping on the source of energy for actions. Again, the annular theory is the only one which will meet this staggering demand — it is the philosophy of sudden downfalls of snow and other annular matter, which brought this age to its close. This characterized each age during development. A new ocean is formed with each age. Only an avalanche from on high could cause such deposits as we now see with rock displacements. May we reason thus, if the sun shining through the clear air of today is powerless to melt ice caps on the Alps or Himalayas, how could it have in ages gone by? If ice dissolution by solar beam is not possible through the clear atmosphere of today, we must look to the aid of terrestrial ringfalls to produce this overwhelming cause.

The study and identifying of the geological ages has not been an easy task. Complete accuracy has not always been possible. The stratigraphic column must be added to and corrected. The geologic evidence sometimes lack conformity. It is often conglomerate. This seems at times to be a puzzle with no solution. Fossils, microfossils, marine fossils, and dinosaur fossils are often found at much higher elevations than where they lived. These too may be conglomerate at times. Some eocene, miocene, and pliocene fossils, may be packed in countless numbers in a common burial ground. It appears an appalling flood must have washed these creatures into a common burial ground. It is possible that some dinosaur fossils as well as others may have been washed miles from where they lived and to much higher elevations, yes, and sometimes to a lower elevation. Annular theory is the only answer to geological conglomeration and disturbance in the geologic strata as well as fossilization blending. I hope I have answered a few of your questions and cast some new light on an old subject. I beg the geologists and others concerned to help fill in the details on this overpowering subject, by first recognizing the annular system of world development. Again, may I say, the buildup of the geological ages, with all of their inter-related variations, and elemental segregation, with its coal and oil, could only have been accomplished by annular process. All others use too puny an agent. Unconformability and conglomerate fossil and geological strata are no longer a mystery with annular theory. They are just as they should be found.

The mineral and elemental distribution throughout the earth's crust presents an unexplainable problem to old-school geology. Let's just consider one mineral, gold, and its distribution in the earth's crust. How could erosion and sedimentation distribute the element in such wide areas of the earth? Gold nuggets are found on the surface. Gold is found in streams of water and mined from them. Gold is found in layers and seams of the earth. Some is as deep as a mile and so thinly distributed it takes a ton of earth to produce a very small amount of gold. Gold is an earth wide product, used by man from the beginning, and still is a monetary material. Gold is on the summits of mountain peaks. Gold may be found on Alaskan glaciers. This should tell us some of it fell in more recent times. Surely as those primeval vapors arose from a molten earth they were laden with gold vapors, and they rode together for ages in the lofty skies, some until more recent geological times. The distribution of gold is universal on earth as most minerals are. A single simple erosion and sedimentation process could not supply the wide distribution of gold and other minerals on earth as they now are.

CHAPTER 11
IS COAL A VEGETABLE PRODUCT?

Because geologists have made this claim for many years it is imperative that the issue be resolved beyond a reasonable doubt.

After many years of study and research of this subject I will use information from Isaac N. Vail, James D. Dana, (1813-1859 U.S. geologist) Professor Andrews, Sir Archibald Geikie (1835-1924), Dr. Anderson, Professor Dawson, and many others including modern day theorists, as well as my own findings.

Consider that we must reverse the law of fossilization in order to conceive of any stratum itself made out of the fossils it contains, it is scarcely possible that a coal stratum can be a vegetable product. (Vail)

We must consider how many feet of vegetation it would require to make one single foot of coal carbon. Some coal veins are 300 feet thick accumulative.

In order to get a new idea accepted it is usually necessary to completely destroy the old accepted theory. The distribution of coal in the earth will not support vegetarian geology. Coal, itself, which is a carbon, disproves the vegetation theory. There are hundreds of ways to disprove the vegetarian theory, yet, only a few are needed to upset accepted carbon geology. It seems obvious, coal seams and beds were developed by some kind of flotation process. Some questions old-school geology has never satisfactorily answered. Why are boulders often found in coal seams if vegetation built them? Why are glaciated pebbles sometimes found in a coal seam? Why are trees and sometimes a small reed found rooted under a coal seam and sometimes extending through the seam? If vegetation made coal why are there so many kinds, and why do we find the occurrence of bituminous patches of coal in anthracite beds? Or we may find fragmentary patches of coal in silurian and devonian. How can vegetarian theory reconcile these inconsistencies? The distribution is wrong.

If coal beds formed in conglomerate and during a glacial period, when did vegetation grow and supply the material for coal? In places such as Ohio there are coal seams a few inches to twelve feet thick. Dana estimated it would take one hundred feet of vegetation to make a coal seam twelve feet thick. (It may require more.) In places, my source of information says, there are as many as seventy different seams, one of them thirty-five feet thick. I have heard of three hundred feet thick accumulative. What about coal partings, some not thicker than paper, some an eighth inch or half inch, extending over thousands of square miles? Why are there these clay partings if vegetation built the coal bed? No man has ever accounted for the mysterious parting features by current theory without opposing all law and reason. If a root bed formed the lower coal, where is the root bed that formed the upper bench? Yes, coal distribution as with other minerals will completely destroy vegetarian geology. More than one hundred seams of coal are found in England according to my source. Another consideration of the annular system with rings, bands and belts, we see carbonaceous material falling to the earth as a great installment of carbon, and each must float away and be deposited in the sea or where natural laws and force placed it. All aqueous vapor of the igneous age must have more or less been mixed with carbonaceous matter. We must include graphite as part of the metalliferous bed of the earth. It is true, superaerial carbon and other matter, must by law, fall largely in the polar regions, (as ice did) coal and oil beds increase toward the north, this will apply to Antarctic as well.

This is not a book for only geologists and scientists. It is my intent that those not well informed on geology and other subjects may read and understand this theme. Most of it is simple law or how could I write it? In distribution, let's consider another simple fact about peat. Many claim this is a vegetable product. Let's see if it is. Because a so-called peat vegetation called Sphagnum, grows in and around swamps and ponds where peat is found, it is claimed the vegetation made the peat. The answer is very simple. If sphagnum moss characteristic growth is only where peat is forming, this should tell us to look beyond the era of the vegetation for its origin. It can be seen by anyone if no peat (carbonite) had previously been deposited where the plant now grows, it never would have grown there. In other words, if the carbon beds necessary to sustain the sphagnous moss were not previously placed there, the sphagnum would not be there. Again, science has things in reverse order. This is true of anything on earth, its food supply must exist before it does. Science often claims the millepore built the food supply it lives from. The absolute reverse is true. If lime had not previously existed in the seas as food, the millepore and many organisms never would have lived.

Going back to distribution again, when we find a thin layer of vegetable carbon in a solid coal bed, it is an accumulation of vegetable debris carbonized by its surroundings. This would also be true of a tar sand such as in Northern Alberta, Canada. A close examination of plant or animal debris carbonized in a coal bed, will show it is often scarcely combustible. This has been greatly misapprehended by geologists in considering the question of coal. Adding to the dilemma of vegetarian geology, I heard a newscast wherein it is claimed there is enough coal left in the earth to last six hundred more years. I will say at a normal rate of use there is enough coal to last longer than six hundred years. That is a lot of vegetation, if they still want to say coal is vegetation.

If coal is a vegetable product, we must ask where all of the coal carbon was before trees developed it and placed it deep in the earth, knowing full well trees cannot make carbon out of thin air? Where was its source of carbon to build from? The tree has carbon within itself, from what source when they say there are none around before the tree? Where did carbon come from? We may ask, where did gold and all other minerals come from, or the ground we walk on? They must have all been part of the original primitive earth. Did each metamorphose form something else? Would a world of swamps and vegetation amid a slow puny distillation give rise to the great vast accumulation of carbon fuel as now exists? Is it reasonable to accept the view that coal is vegetable carbon changed to a hydro-carbon, and subsequently partially changed to oxidized hydro-carbon?

The piercing light of philosophic truth will deny this theory so widely accepted. Coal seams are boundless and vast in extent, we are unavoidably bound to the conclusion that a process a thousand-fold more stupendous and competent would be required to produce all of the forms of carbon now found in the crust of the earth. A great primitive furnace, a boiling, burning smoking planet, a world-furnace is the only competent means there could be to distill these vast carbons and other minerals and build a stratified world. A world gas retort must have distilled all of the carbons of the earth. It is obvious that this process could never have obtained, if the foundation had not been precisely laid and elements previously suppled by the former. We see fuels formed in immeasurable quantities by the only process commensurate to in creative power to do so. The only possible terrestrial source was the aqueous sediments of the original earth. Why do scientists ignore law and claim some mysterious plant food process developed the coal of the earth; when the process is a non-supporter of animal life; a non-supporter of combustion; transformed into a solid carbon, and through some secondary combustion; transformed by accident into a hydro-carbon. Why force a puny process when the carbon is already at hand?

PERMANENT FOUNDATION

We must build on a foundation a thousand times more permanent. No man of mature reflections will deny the earth was once in a fiery molten condition. The original earth in such condition contained all of the forms of carbon in its rocky frame would be driven outward

and form annular rings around itself, along with vapors; positions
would be determined by their specific gravities. It is axiomatic law that
these carbons and aqueous matter would form segregated and ag-
gregated annular systems. Upon declination, these rings would over-
canopy the earth with belts falling, and the oil and coal is right where
it was placed by this system. We may assume that all worlds are made
alike; therefore Saturn and Jupiter must have carboniferous strata
revolving about them as primitive carbon distillations thrown there
when they were hot and igneous. Why shouldn't carbon distillation be
part of the economy of world-making process? Contrary to the opinion of geologists, carbon was formed without the aid of vegetation, it
was supplied directly by the primitive process of the igneous earth.

If geologists claim coal is a vegetable product graphite also must
have had a vegetable origin. Now if they say animal may have aided
the process this makes it even more difficult; since it is the carbon that
makes the organism, not the organism the carbon. Graphite is found
in the oldest sedimentary rocks according to my source. Now if plant
fossils are ever found in a graphite bed this would not mean they made
the graphite. They would become graphite merely because they are imbedded in graphite; just as a plant form falling in sand would have
become silicous fossil.

Aren't we forced to accept the miraculous, a standing tree while
vegetable debris accumulates to a thickness of 40 feet at its base! How
could this tree continue to grow in such a swamp? An important
feature greatly misapprehended by geologists that vegetable carbon in
a bituminous bed is not bituminous. The accumulation is just carbonized by its surroundings. Much more could be said on this point;
law would demand that coals necessitates that the vegetation should
contain the same elements as the coal it is supposed to have built.
Fossil vegetation speaks plainly in opposition to the vegetable origin
of coal. Vegetation so abundant in the clay above the coal seam are
foreign matter in the clay, simply an involved vegetation, just as in
coal. The idea that plants and plant remains are found in abundance
in coal seams is misleading; when the truth is in some coal veins they
may be very scarce or non-existent. A simple fact that there are fewer
fossils remains in some coal than in the superimposed beds, shows that
the carbon occupied less time in collecting and forming into a bed. A
flotation factor must have built coal beds at times very rapidly; fossils
were just caught in the process.

Fossils in coal are found horizontal as well as often vertical. It
seems coal is not metamorphosed peat. Try to imagine a world of coal
marshes cover 10,000-100,000 square miles, all about the earth. In
order that a second coal seam should form after 20, 50 or 100 feet of

clay, sand and lime accumulated over the buried carbon bed is again turning to the miraculous. Now if we are to say these developing coal beds sank into the sea and arose twice this would be remarkable. When as in places many coal seams exist accumulative to 300 feet thick and to a large extent this must be happening simultaneously in all lands we must add miracle to miracle, with our credulity unduly stretched. Are we to admit this oscillation of sea and land must repeat 50 to 100 times to form these coal beds and that it must at times lose its regularity, in that some coal beds are at abyssal depth and accumulated limestone, isn't this an absurdity? How did the beds of other lands join simultaneously in this process? It seems that the coal is planted immediately upon an aqueous-formed stratum, and all the beds between coal beds are also aqueous strata, and we must find some primitive carbons as aqueous formation in the midst of sedimentary beds. Submergence and re-elevation of beds hundreds of times is not a true concept. If continents and ocean were constantly oscillating and trading places, the strata of the world should be much different than it is. We have 10,000 freshwater lakes alone in Minnesota, and other states too. A comprehensive glance at the difficulties we find the vegetarian theory to be unnatural. A half-way explanation will not do.

Dana, in referring to the grand structure lines and frame-work of the continents, is forced to say: *"There is strong reason for concluding that the continents have always been continents; and that while portions may have at times been submerged some thousand feet the continents have never changed places with the oceans."* The truth derived from the carboniferous conglomerate beds, it is a fact that they were sychronously formed the world over. In my case I do not oppose the idea of local submergence, on a small scale such things do naturally occur. Next, it is obvious the whole world could not have been flooded except by a downrush of super-aerial waters. It is a matter of fact that there is no such thing as parallel coal viens according to my research material. (Professor Newberry) They may appear to be but on a worldwide scale they are not. These beds are evidently lying as they were placed. The annular theory would make both an aqueous and sediment bed. To the consternation of the vegetarian geologist a large coal seam may be made up of two very different kinds of coal, bituminous and another with a separation no thicker than writing paper. Sometimes these veins trade places in nature with the above, and again the first one will be below. Why would slow vegetive growth over millions of years do this act of miraculous?

Vail said; *"We must conceive this carbon as susceptible of transportation and change as any other sediment, and local beds of fine carbonaceous mud, which the cannel carbon really is, could not avoid formation while currents ran, any more than sand or clay. Admit all such beds to be sedimentary deposits of annular carbon and every mystery in their formation vanishes.''* Carbonaceous shales, so universally prevalent, must have had the same origin. A carbon bed was formed, currents fed it material to cover part of that bed, another supply of carbon settles down on this and the process repeats itself building the coal beds of the world. A coal vein made historic by Charles Lyell, had seven vertical veins which all joined together in five miles. To say this coal bed submerged seven different times to form these beds is very unphilosophic. The simple answer is they are just sedimentary beds.

Let's introduce another obstacle to the vegetation theory. It is a well known fact that boulders and rocks foreign to the localities of where found are in some coal seams. A boulder with a weight of over two hundred pounds was once found in an Ohio coal seam, which had marks of glaciation. The boulder was in the middle of the seam so it shows the bed finished forming after the boulder was placed in the seam. How do rocks and boulders get into coal veins is the question? How could vegetarian geology place them there? In order for these boulders to arrive in its new surroundings it must have been floated over the surface and dropped in the coal deposits. Sometimes water worn pebbles are also found in coal veins. Yet, there must be hundreds of coal deposits around the earth never inspected by the eye as yet. Where are we to fit the vegetation theory in all of this fact?

Another evidence I have mentioned before quite as positive as the boulder problem; that of coal partings; that have divided coal seams not once but many times, and sometimes a few inches of clay or as little as the thickness of paper. There are many coal seams in the U.S. as well as other countries which have these persistent coal separations, in varying thickness. These coal partings in varying thickness may extend over thousands of square miles, some may have a few vegetable fossils where thick enough, though many are barren reaches of clay. Why are there never enough if any fossils in or below or above these clay partings to form a coal bed? If men are not bound by antiquated opinion they must see that both the boulder and coal parting in coal seams is an obstacle they cannot remove. Other countries of the world have reported similar findings in coal veins. It is world wide. I do not know to what extent microscopic life may exist in coal seams if at all, though this will not give support to vegetation geology.

Why do most coal beds have a clay underlayment? The persistency of these many peculiarities in all lands, leads us to look at the annular system as its source. When we see trees standing in and surrounded by this clay, and rising through the coal seam, and even penetrating many feet into the overlying rock, we are forced to admit a rapid accumulation. Every standing tree in such position is not convincing to the claim of vegetable distillation either in coal or clay. Then add immense periods of time and we have insuperable evidence against the vegetation theory. The beds had to accumulate during the lifetime of the tree. Not only the coal but the sand accumulated around the tree. How could the tree survive such an ordeal? These trees growing in its own swamp bed must have perished when involved with ocean augmentation and annular carbon deposits. This gives us assorted beds and strata. This is an extremely rich field of thought and I cannot begin to bring you all of the information from my many well established sources of information. I will add one more point though, that it is self-evident that peat vegetation could never have assumed a foothold in any region if the peat foundation had not been previously laid down. If I had space observations by Vail, Dr. Anderson, and Geikie would amaze you. Plants did not form peat bogs, they live from them.

If vegetarian geology is true shouldn't the greatest coal beds be found in the tropics? Peat and coal are not found where the vegetarian wants to find them. They are found where annular theory places them. As Dana estimates, a peat bed 30 to 40 feet thick would require 240 to 320 feet of vegetable growth. Did vegetation fill these swamps with carbon, so it in turn might plant itself in a swamp to fill the lakes with carbon? Some have suggested that if all of the coal buried in the earth were moved to the surface it would make a layer of coal ten feet thick all the way around the entire earth. Others say more and I would agree, it would be more. Some experts estimate it would take at least 12 feet of tree vegetation to make one foot of coal. Other expert geologists say it would take a lot more than 12 feet of vegetation to make a foot of coal. If we take the low figure of 10 feet of coal around the entire globe as an estimate of all the coal buried in the ground, we would need 120 feet of tree or vegetable matter to make this coal. I agree both figures must be increased. Nevertheless, in an area with 300 feet thick coal veins accumulative at the low figure that would be 3,600 feet of vegetable matter to form such a vein. One expert estimates that deep coal beds 100 to 120 feet thick would require 800 to 960 feet of vegetable growth to build them. It is pure logic that no matter how luxuriant a plant growth prevailed there just is no way vegetation could have placed the coal beds of the world that are now known and yet undiscovered. One expert scientist says, *"It would require 60 feet of compacted plant debris to form one foot of coal."*

Inconsistencies meet the vegetarian geologist at every turn. May we now ask why vegetation would develop so many different coals such as anthracite, bituminous, lignite, and a dozen variations in between? Too, there is the occurrence of bituminous patches of coal in anthracite fields. Further, to stumble the vegetarian geologist — the old theory, there is the occurrence of heavy anthracite beds, where, according to old theory, bituminous should prevail. One report shows that *"fuel carbons and continental snows, with frequent occurrence of conglomerates, either directly below or above coal, are frequent associates."* This is inconsistent with the vegetation theory. To build coal beds in glaciation then, is not in harmony with vegetation growth. Coal has been found in fragmentary patches where it should not be found.

METAMORPHISM

As my source says; *"it has become an accepted belief that all anthracite and other hard forms of carbon found in the earth's crust are metamorphosed beds of soft carbon."* Is this deduction a logical one? How is it then, that vast coal fields planted in aqueous crust miles from any igneous agency, are now in anthracite or semianthracite state? Now it seems to be a fact, that metamorphism is an overworked theory in the entire spectrum of science today, as well as the past. The igneous earth was the distillery of the carbons it had control of. Are we to accept the claim that the last puny fires of the earth metamorphosed all this fund of carbon, when the primitive fires of the incandescent earth and measureless oceans of carbon did not? Anthracite and all hard carbons exist in the earth's crust as an inevitable product of that igneous process. The philospher should be impelled in spite of education and prejudice, that very little metamorphism has occurred in the coal carbons of the earth. They are a carbon unchanged since placement.

How could bituminous change to hard coal by merely driving off the volatile constituents? I wish that I had ten pages to bring the figures I have before me to you. A mining engineer once showed me the ash content figures of many coals, and it does not support the metamorphism of bituminous coal to hard coal. Dana, too, has some figures on this ash test, as does Vail. Using 100 pounds of bituminous coal subject to enough heat to drive off 20% of ash does not come out in favor of weight to become metamorphosed hard coal. Perhaps sometime I will publish these statistical facts gathered from my expert sources. I will say this, that ash figures and samples may be misleading unless we use many available from all over the coal fields of the U.S.A. In going over these figures a geologist will drop the doctrine of metamorphism in reference to anthracite, and admit it is as much a normal product as bituminous. Men better qualified than I am could gather information from coal literature of the world, and see for themselves that anthracite coal is not metamorphosed from bituminous coal. It requires the agency of excessive heat to make anthracite; every evidence is that this heat was brought to bear upon this coal before it was laid down, not after. Then we must admit that such coals are the primitive products of the igneous earth. This is only a demand of physical law. If we should find an increase in coal fossils in an area, generally the fossils in the clay and rock increase even greater.

I will say again as Professor Vail did; that fossils in coal are generally mineralized charcoal, and difficult of combustion. We are forced to admit the plant is foreign material mineralized by their surroundings. We must look beyond the plant for the origin of the bed. A further point is the charcoal found in a coal seam as fossil material frequently analyzes from one to one-and-a-quarter percent, and sometimes as low as three-quarters of one percent. That is, the part of a coal seam known to be vegetation is so free of ash as to argue the inconsistency of claiming that the whole bed is a vegetation. This is according to my expert source then, and if the ash content of the main coal bed is much higher we know the fossils did not make the coal bed.

Because I have believed since a boy the idea of ocean augmentation I will give a direct quote from Isaac Vail, made well over eighty years ago. We as well as all geologists are aware of how much coal there is in England and Europe as well as vast horrendous amounts of oil in the North Sea and beyond, Newfoundland and at Hibernia, much undiscovered as yet. Quote from Vail; *"Turning now to the eastern border, (U.S.A.) and finding no tertiary coals there, we are led to believe that a narrow continent stretches from America to Europe, across the present bed of the Atlantic, thus hindering the southern flow of carbon along the Atlantic sea-board. It is now fully conceded by geologists that through the teritary epochs such an isthmus of land reached from Newfoundland to the shores of Europe; if this be true, what a fund of tertiary carbon must lie at the bottom of the North Atlantic!"* We know the rest of the story. Could vegetation and some fish develop all of this magnitude of carbon? Vail said further, *"And again, since the absence of these coals on the eastern border of the continent forces the annular theory to demand a barrier across the flood ground from the north, why did this barrier come just where and just when it was needed to support the theory? These, it is true, are minor links of evidence, but they are links none the less."* This is not such a minor evidence now, is it?

Every reader must comprehend that if the vast lignitic coals is a vegetable production, it was present in the tertiary atmosphere as a deadly poison. Look at the immensity of the coal fields known by geologists, and conceive what an atmosphere that was. A look at the cretaceous seas and behold them filled with breathing animals; yet vegetarian geology would give us an atmosphere with the highest degree of destruction to life. How can we reconcile this? What a tribulated path the vegetarian geologist must lead! How could there be a coal-forming age, when life was abounding, this means the defiance of law, if we admit coal is a vegetable product. In the last 5,000 years how much coal has plant life made? The theory that fails in one point is a complete failure. Metamorphism, having failed in the most essential particular, is a complete failure.

There is an endless amount of information on these subjects from a hundred sources I have at my command. I feel I have short changed you. The earth was subject to such a destructive distillation during the igneous era; and, therefore, these forms of carbon were placed in the earth's crust, and placed there after the primitive fires died out. There is not so much as a feature connected with the formation of coal that is not readily explainable by the primitive carbon theory, and annular declination. Thus all ages were more or less characterized by carbon falls, and no age could be exclusively carboniferous.

As Vail said, *"I will say to all geologists: Come to this new field! The vegetation theory cannot be true! You all know full well that these stubborn facts are continually multiplying. How many years must roll by before you recognize them and come to grips with them? Just as with oil, it is well known that the bitumen in its rocky matrix, when subjected to sufficient heat, is driven out as an educt, not as a product of the rock. Being an educt, we must look beyond the rock for its origin."*

Before I let this matter rest, I must make a quote from Vail as follows: Quote: *"Now let us imagine a world composed exclusively of water and sandstone. Let it be fused or melted to its inmost depths by inveterate heat, as millions of worlds are today. By this heat its waters would be vaporized and driven away from the fiery mass, and the core would be a mass of melted silica, and an atmosphere of aqueous vapors would surround it. Now, if the heat be increased so as to make the mass a shining sun or beaming star, the silica would be vaporized and also driven away and made to commingle with the watery vapors, and if the melted mass contained limestone, iron or lead, these substances would also be vaporized and the vaporous atmosphere would contain gaseous matter of all these substances. And it must be seen that in the universe of law, as the mass becomes cold, these vaporized elements would condense, in order of their susceptibility of fusion and vaporization, and, falling to a common center, would form a spherical mass, not of water and sandstone as before, but one com-*

posed of all of these elements. It must also be seen that the aqueous vapors would be the last to condense, and, moreover, the last to fall from the position which they must have taken under the reign of heat, repelling from a focus; and, while all these materials thus vaporized and afterwards condensed, must to some extent become commingled and form, and form just as we see under our feet, a heterogeneous world, yet there would be on the whole some definite and regular order of strata arrangement. For, in the condensation and consequent precipitation the heaviest and most refractory minerals — minerals most difficult to fuse and vaporize — would separate from the rest and settle first. Beds of silex; almost pure, and silicious beds containing iron, calcium and every other metal or mineral contained in the fiery envelope would recur in some kind of order in our hypothetic world. There would be beds of metals arising from this fiery distillation nearly pure, under the law of elemental assortment and segregation. We see this law abundantly and universally exemplified in the entire structure of the earth's crust." End of quote.

My contemporaries have contributed much to my book, yet I have relied heavily on the factual accurate information of Issac Vail. The above describes so well the world we know to exist under our feet. A world building process must account for elemental segregation, separation, and assortment, in the distribution of its mineral wealth. Any postulated process which does not, is not valid. This has been a major stumbling block.

Another point I must make: oil, gas and carbon content of the Antarctic region. It has been my claim for many years huge fantastic amounts of oil and gas exist north of the Arctic Circle. Drilling has now proven this to be true. Then I must also say, vast unheard of amounts of petroleum, and coal, are buried in the Antarctic region of the world, by the same annular system of geology that placed primitive carbons in the Arctic. The big question is, will it ever be found and will it be recoverable? The cost will be great.

CHAPTER 12

CIRCUMPLANETARY DEVELOPMENT

In considering the earth's steady progress of development to this point, we must use known physical laws. Most are simple, not complicated. For example rotation law cannot continually violate axial law. Therefore, regardless of the shape of a rotating mass, it must obey axial law, or rotation dynamics. We may feel only circular objects have an axis. Any object that rotates must find an axis, that is, as much material on either side of axial as the other. The eccentric rings on Saturn do not violate orbital law in totality of rotating mass, it must obey axial law, or seek axial law. If not, it will ultimately come apart and seek new positions in harmony with law. Nothing makes up its own rules though it may appear to. We may assume the earth now has as much mass on either side of axis as the other. If not, it will adjust to axial law in automatic obedience to physical law. In the planets, we may have outside forces of gravity affecting them, though this is another subject.

In orbital and axial law, distortion of any major degree cannot sustain without ultimate retribution, which is correction or destruction. The original igneous earth found axial and orbital position, and still has it for the most part. The spinning earth in geological development was fed material and water to it from an annular ring system whirled around itself from its own material, which would, of necessity, obey these fundamental laws. Some may argue, that if great mountains were heaved up, as is the case, this would throw the earth out of balance with axial law. Not so, because mountains were heaved up on all continents for an equal balance. Then too, as the annular theory proves, the earth was continually augmented with water until the last great downfall; we all know water cannot violate axial law, rather it would stabilize the earth. This is proven by a tide each day caused by moon attraction. Gravity and rotation will hold even a water peripheral boundary constant as an average, to axial. Polar ice caps do not violate these permanent universal laws. Polar ice caps were supplied to the poles from this earthly ring system, mainly the last watery ring to fall which also changed the climate of the earth. When large areas of ice took possession of polar regions of the earth, much of the ice displaced water, which would automatically obey axial law. Where then, do we find any distortion? The spinning earth started out with axial equilibrium and still has it. In millions of years of development, as it whirled material into orbiting rings around itself, later to reclaim this material, as pouring water on a spinning wheel, where do we see violation of axial law? Material would return in stability. It developed as a turning wheel or potter's table, it still turns. If, at times, the earth did not have axial stability, it would seek it.

There just is no place for the theory of continental drift in the laws of true physical law. Once the arch of a continent began to form, it always remained as such. Some scientists claim the earth's crust is continually in the process of rising and falling, alternately becoming ocean bottom, then continent and mountain peak, cannot be correct as this violates all known physical law. I will agree that Greenland was once tropical, yes, the entire earth was once tropical, before the last great ring of water fell to the earth's surface, which would include ice at the poles. Then there are those who claim inconceivable tonnage of glacier ice accumulated in lopsided deposits, throwing the earth's axial equilibrium off so it shook the continents into drifting. Please explain where such great amounts of ice came from that it would cause centrifugal abberations of the earth?

Could normal weather patterns form this ice, knowing full well no such weather patterns could exist under such conditions? It requires solar energy which is heat, to build a weather pattern before snows may form. Gravital, centrifugal, orbital, and axial law, has operated smoothly and constantly throughout all planetary development. Nobody knows what gravity is, nobody knows what the independent inertia source is of a rotating planet or satellite, or for moving planets around a primary such as the sun. Even more strange, is that some planets and moons may not rotate at all. Carrying this a step further, some satellites on a primary may rotate in the opposite direction of its neighbor or in at least one case, a moonlet has a reverse orbit to other moons on the some planet. Gyroscopic rotation is not needed to hold satellites in a permanent gravital position with its neighbors in the solar system. Saturn has one moon which travels in the opposite direction in the Saturn gravital field. Where in physical law do we find any rule which says it should not do this? We do not know the original circumstance which put all of these actions into motion, do we? Who would dare say there is some violation of physical law, orbital or otherwise? Perhaps the most disgusting thing about the universe is, that it is governed by such simple laws even a child may understand them. Take a short piece of string and tie a small weight on the end, then swing it around in a circle, as we often did when we were children. Herein is emobdied many of the basic laws of the known universe.

Just as nobody knows what gravity is, or the inertia for revolving and rotating planets, the 23½ degree annual tilt of the earth's axis each year is not known. This tilt seems to violate axial gyroscopic law. Yet, just because we do not understand what causes a known action, does not necessarily make it a lawbreaker. Isn't it true, that much of what we have been taught in historical times, even in our lifetime, was not true or correct, and often the opposite is true? It would seem simplicity is the greatest stumbling block of modern technology. The more that is learned, the less we know and more mistakes are made. An example: Astronomers have found much about big black holes in space. According to the New York Times, astronomers have found the largest hole yet. Completely empty. Yet, in the "big bang" theory, the distribution of matter and motion of the universe on the average is homogeneous in all directions. This "giant hole" is just another unexplained phenomenon for science to ponder and try to explain. I am not critical of research into the unknown, some is necessary. It seems far too much time is spent on the unknown when perhaps more answers would be found if more were spent on the known.

This brings us to the subject of the earth in its annual orbital trek around the sun. Some may say here is an example of orbital violation in that the earth travels in an oblong, eccentric, elliptical orbit around the sun. Again, we must consider planets are affected by both internal and external forces in their travels. Accountability may be in the fact that a satellite always travels faster when closer to its primary, as law would demand. We cannot measure all of the actions of the solar system; there is velocity, attraction, and densification and many others. We do not know all of the unknown forces which come into play. If we knew them all, would there be found any distortion in the workings of the solar system and its movements? I have always thought unknown planets, stars, or blackholes, or other factors not known as yet, are exerting forces unknown as yet. This is not difficult to understand. There must be accountability somewhere. If there is distortion or appears to be distortion, it must be compensated for by other forces known or unknown, or the system would come apart. This is just logic and reason. The solar system must be drifting through space in unison with the Universe.

There has been and will be, ring system declination (rings on planets descending to the surface) within the solar system. Compensatory law again comes into play. In this case, planetary volume would increase along with desification — total volume would not — so gravital balance would be maintained within the system. There would be only a small proximity change with other planets with ring declination. If there were a small loss in gravital balance and attraction loss due to proximity change, new planetary position would automatically correct and compensate for a balanced system again. Keep in mind, planetary ring declination would be slow over millions of years so adjustment and correction within the system would also be very slow. Balance would continuously be maintained within the system to the salvation of us all.

Perhaps we are getting deeper now than we care to. We may ask, what are the constancy factors in solar law? Are not its functions and actions governed within itself for the most part? Has there been a substance loss or gain in its lifetime til now? The Master Plan, built in compensatory law. The question comes up again, what is the independent inertia for rotating and revolving orbits of planets? As a boy I spent hours with a gyroscope spun by pulling a string. I often wondered why it would slow and stop turning after the initial thrust. Planets do not. If you ever find the answer, please let me know. I had a writing pad on my easychair while I wrote the above few paragraphs in a relaxed mood watching the Lawrence Welk Show and the Mandrell Sisters.

I do not mean to imply that the earth may not, at times, be a small degree out of line with true axial rotation, now or in the past. Nobody knows though it would not be impossible. Tides each day may affect this action in a way we do not understand. Perhaps we may assume, in a twenty four hour period, even with tidal change, the total volume of the earth will average near true axial rotation. In a mechanical device, diameter and velocity will rule precisely in axial law. Balance must be perfect in very high rotation. This is very apparent in a high performance jet engine. While flying copilot with my friend in Palm Springs, California, in a twin engine Turboprop plane, he explained that some jet engines have an rpm in excess of 40,000. In this case, no distortion of axial law could be tolerated or the engine would explode instantly. This will give us some idea of two extremes. The slower a mass rotates, the less it must obey axial law. In the case of satellite, it would still obey orbital law though not have axial rotation.

We should be thankful that all physical law is compensatory. This was a lesson I learned when very young in flying small, two seater aircraft. At that time, nine out of ten landings we made were off airports in the rural areas. It took a great deal in ingenuity to land an aircraft with a weight of less than one thousand pounds on a country road with a twenty mile per hour cross wind and telephone poles on both sides of the road. Yet, we did such things without mishap. This was not dangerous as we knew exactly what we must do. Physical law was our guide. The only flight instruments we had were a compass, nonsensitive altimeter, and engine tachometer (rpm), and our speed indicator driven by air moving over the aircraft. The last gauge was our lifeline and in final landing was the only one we needed. This speed gauge linked us to air motion, not to ground speed. The help from thousands of hours of experience and natural feel was most welcome. Fancy instruments used in modern high performance aircraft would have been useless to us at that time. A keen sense of compensatory law was our total regulator.

This, however, is true in high technology as well. May we never forget this fact. As time and progress developed, we had planes with electronic and gyroscopic instruments. A simple electric or air driven gyroscope instrument for turn and bank and blind horizon is still the main instrument of flight and is used right up through space travel.

My point is, no matter how complicated an objective may be, it originates in simple rules and laws. To make a space flight to the moon or any other selected point, original accurate trajectory (a curve that passes through all curves of a given family) must be established. Other factors will fall into place. Science has always fascinated me even as a boy. Yet I see no reason to make a god out of it. It is nothing more than a natural consequence of things and actions set in motion long ago. Many are still arguing the issue, by whom? If they cannot resolve this issue how can they resolve the others?

LAWS OF UNIVERSAL PLANETARY EVOLUTION

How did the rings on Saturn, Jupiter, Mars, and Uranus originate? There are many detailed theories available to read in many publications, news magazines, and books. They are interesting and educational. It will be my firm position that planetary appendages, rings on planets, originated as material from the planet itself. Not from outside accumulation of matter, though a small amount could have. My mind is not closed on the matter and should information to the contrary be forthcoming, I will change my stand. There are alternative explanations as to how the rings arose. As I have already stated, the earth once had its own ring system, now descended to the earth and is again part of the earth.

Rings on planets are detached disks of matter arrayed in the equatorial plane of the planets. They are held there by gravital and centrifugal forces caused by their velocity, speed of rotation. The field is magnetic, gravital, and centrifugal, there is a constancy. This supports annular theory. I do not claim each planet is representative of the others. There are varying circumstances, yet there are planetary conditions that must be alike in every planet. The planets were born in great heat and igneous flaming fire, along with rotation. Carbons were distilled and along with other minerals and elements, were driven far from the firey center to float as infinitesimal material in the solar heavens. Unconsumed carbon is a prominent component of this system. Rings are located according to unchanging law, at a distance from their primaries measured by velocities and gravital force. It is a mathematical and philosophic necessity, an analogous fact, that sister worlds to our own, must be ringed and belted. We have our own satellite as a model. We cannot conceive of such enormous area of sea bottom, should even a single foot of sink take place, what would happen to shore lines and river outlets, without replacement of water by ocean augmentation.

Do not the rings of Saturn, Jupiter, and Uranus share

similarities? Each has moons. Water ice is a major part of the ring systems as it was so on earth. On Saturn there are deviations from circularity in its ring system, which is as it should be with moons and other factors causing forces to vary. Also, Saturn's period of orbiting material matches its rotation. Its magnetic field rotates with the planet. The architecture of a planetary ring system is very interesting. There are moons outside of the rings and moonlets embedded in the rings with gravitational forces and other forces. Everything in the ring system has a common orbital motion; they travel in the direction of the planet's rotation. From reading scientific articles, it seems there is a slow energy loss due to interacting forces within the system. This may indicate the appendages will in future time, descend to the planet as those on earth have done. If enough distortion and energy loss develops in the ring system by interacting forces, the rings will trend lower with increasing density and belting, they may fall to the planet. This is not to say it must happen. Interacting forces of rings and moons is not fully understood. The magnetic field of Jupiter is a few degrees tilted from its axial rotation.

I am basing many of my remarks on intelligent facts as they are gathered from space missions and other sources of information. Much of this information on the planets tends to support my theme of ring development on earth as well as sister planets. An example, it is claimed many of the planets were once larger than they now are. This is as it should be. If the material in any planet became the material for its ring system, it would have been larger before it lost material to its rings. Likewise, should the rings descend back to the planet, it would again be larger, as it originally was. It is felt great heat was associated with a belt system which belts on some planets have disappeared. Where could these circumplanetary envelopes go to? Why should gravity give them up? It would take new energy. It is suggested friction within the envelopes may have caused them to fall to the planets. It is also believed the planets rotated faster during development increasing centrifugal force. If so, it is reasonable that the belts or envelopes moved farther out as physical law would dictate. The first belts would become the rings we see today. On descent, they would again become belts, later to fall to the planet. Formation factors then ended and the process reverses — millions of years later. Many qualified scientists agree the giant planets have had a heat and energy loss, so it only seems reasonable they will fall back into themselves. They cannot expand further without new energy.

There are rival hypothesis as to how circumplanetary development originated. The interesting thing about it all is most scientists agree that the planets and moons are made up of ice water, rock, other minerals, and gas. This I will agree with and is as it should be in my annular theme of geology. This still allows for variation in each planetary system. One theory is that the primordial ring material came about when a large body came into the planetary area and fragmented causing the rings to form from its debris. The agent of fragmentation would be tidal disruption, collision. It would seem tidal disruption would destroy so much energy and velocity, the material would descend to the planet, rather than be thrown into a ring system. A spinning wheel will whirl material out and around itself, as it adds velocity, not diminish, as tidal disruption would reduce. I am sure the ring material arose from the planets themselves for many reasons. Yes, it is true, there could be some tidal disruption by stray asteroidal action, just as the earth has. The velocity loss in such action would tend to bring substance of collision to the surface of the parent body. On earth, the atmospheric friction burns up most falling material. The return of Sky-Lab tells a story we cannot ignore.

Another hypothesis postulates that a single large moon or a series of moons collided catastrophically with a stray meteoroid, and the resulting debris became the rings on a planet. True, craters on some planetary moons seem to indicate collision. The same physical laws will apply and there is a great doubt in my mind such action would suffice to build circumplanetary ring systems. If we lived on another planet and were observing the earth from millions of miles away, we may feel the irregularities of the earth were caused by asteroidal collisions. We know this is not the case. Perhaps there are some meteoroidal collisions within the planetary trajectories of ring systems (not meaning small particles). Material traveling within a given trajectory would not have that much difference in velocity and would approach each other more like two space capsules docking. Even so, if this collision caused velocity and independent energy loss, a tendancy to descend would be the case. An influx of meteoroids passing into the trajectory path of a circumplanetary system, even without collision would create a drag on the system's energy source before a rotation catch up could take place. With collision, the drag would be even greater. Too, there is no way of knowing at what angle these meteoroids and asteroids would approach from. Their capture by a planet must diminish energy force not enhance. If an object entered a planetary system against rotation, the drag would be greater than if it entered with the direction of rotation.

It is thought by many scientists, interplanetary meteoroids and micrometeoroids excavated many craters many times larger than the impacting body. The ejecta is thought to be the material for orbiting ring systems. Again, it goes without saying, in order for such collisions to take place, the objects cannot be traveling in the same direction with the same velocity, or how could they occur? There must be a crushing impact to throw material out of the gravitational field of the planet it impacts with. If the ejecta lacks the energy to escape from the moon or moonlet it originated from, where would it get the energy to start orbiting the parent system? I would agree with one author who says; the moonlets and large rings on planets date back to the early history of the solar systems; they are contemporaries of the moons and planets. I would say, because ring systems always rotate in the direction of the rotation of the parent planet, and from an equatorial plane, they must have originated from the parent planet. Most moons will rotate somewhere near this plane though they may be off some.

Yet to be answered is why one moon on one planet rotates the opposite direction from other moons on the same planet. This is a mystery to us all. Are ring systems static now, or are they still growing? Some feel they are growing. It would seem they have about reached their apex and are ready to recede, though I will leave this door open for now. Many are in various stages of development. The ring systems have been a challenge as many other creative processes have been. There is much information available from scientific research on these subjects we may all learn from. It is hard to contemplate why science will sometimes use such a puny agent to accomplish such grand results. We must have a competent agent equal to the accomplishment. To just add billions of billions of years of time to the action will not compensate. It may detract from credibility of the objective.

CIRCULARITY — A UNIVERSAL LAW

Scientists and astronomers who claim outside asteroidal material built ring systems on planets create many unanswered questions. We only have a few factors to consider some of which are gravital, centrifugal, and velocity. Circumplanetary development and the activation of them is a very interesting subject. How many possibilities do we have? If scientists claim one single massive influx of meteoric or asteroidal matter engulfed a planet forming its ring system we have some monumental problems to answer.

In this case some of the material would enter the planet's gravital field against its rotation and some would enter with the direction of rotation. If the material entering with rotation were captured by the planet's gravity and began to come around the planet as a ring it would collide with material which entered the other side against rotation. Velocity and inertia energy would be destroyed to such an extent the material would descend to the planet and not become a ring system. If such a massive drifting influx of asteroidal material entered a planet's gravital field, in the direction of rotation it would curve around and bounce out the other side and not become a planetary ring. The planet would not have enough gravity to hold the material, and if it had enough gravity the material would join the planet. If such a mass of outside material entered the planet's gravitational field against rotation, there would be only one possibility, and that to join the parent planet not form a ring system.

Next, do these astronomers postulate the ring system on planets were developed from a single influx of material to each ringed planet, or was material slowly delivered to the planets over millions of years? This position is even more troublesome to answer. In both cases there would be a great energy loss causing the rings long ago to descend to the parent planet. In this hypothesis circumplanetary material would have to borrow energy from the parent planet, causing ring declination in short order. Where else would rotation energy come from? Is it ever possible to invert an energy unit without an outside force? An outside force seldom has perpetuity — a steady flow of energy. We still have a factor yet not accounted for. Rings on planets as they now exist always rotate at the equatorial plane. Therefore this stray material could only have entered a planet at the equatorial plane. Who, then, or what, always systematically guided this influx material to this single plane? Are we not stretching chance to the very limit?

We have another problem in this theory which I do not accept. Untold billions of new tonnage would be added to a planet whether by rings or to the planet. Would it not then need to seek a new position in the solar family of planets? Any proximity change would cause all other planets to shift positions. At what stage of solar development was all of this rearranging done? Now then, scientists say, life on earth is billions of years old, the earth will not support life in any other planetary position than it now is, or rotation as it exists now.

Another problem. Let's take a look at the ring system on Saturn. There is elemental segregation in all of its rings. If a single influx of asteroidal material stacked them all up, how did it get elemental segregation in all of its rings? Why is there so much ice water in its outer ring? How and when was all of this segregation done if built from a single influx of material? Then if we take the other position that each ring was a separate influx of material to Saturn, this is the fraction of all law in the universe. The rings would find themselves in reverse order. Where did its water ice come from?

There is only one very simple answer to this very complex problem. The rings on each planet arose from the planet's own material in its igneous development process. In other words, each planet always had the same mass from its beginning (Big Bang if you like). This must be true of the earth's one time ring system. Because the planets had the same amount of mass from their beginning, when rings developed or descended, there never was a need for them to seek new positions in the solar family or system. I do not mean to imply that asteroids or meteorites are never captured by a primary planet. According to my scientific books it is proven that circumplanetary rings always rotate at the equatorial plane of the planet. This should tell us that they arose from the planet. Asteroids and moons may be off the equatorial plane in their movements, though close. It is puzzling that two moons on Neptune crisscross the equator diagonally.

The planetary proximity factor must remain fairly constant in the solar system. In order to calculate the weight of a planet we must include its rings, if it has any, and all of its moons and moonlets and all debris in its possession. This causes planetary order, and rings could rise or fall on planets without ever changing its weight or orbital proximity to the master planet, the sun. There would, however, be a variation in the proximity factor with moons, moonlets, and other debris on each planet when a ring system arose from the planet and returned to the parent planet. Moons, however, could not leave the gravitational field of the parent planet or they would not have been there in the first place. They would simply move closer or farther out with the ring system's rise or fall because of a small gradual attraction loss or gain due to a proximity change. We must assume the laws of gravity, centrifugal, and velocity remain constant.

Some scientists feel there are no absolutes. This leaves many doors open for them to come and go as they see fit. Sometimes I get the impression astronomers and scientists are apologizing for the way things are made because they are not exactly the way they think they should be. There are some variables and we do have some absolutes, or we would not be here and you would not be reading this book. We, as imperfect humans, stand on the very threshold of wisdom and knowledge. Yet, as has been said, we cannot see the forest because the trees are in the way. It is plain, the material in planetary rings arose from the planets themselves in support of the annular theory. It is therefore, not possible to add or take away, large amounts of mass from existing planets without throwing the planets into disarray among themselves in relation to gravital pull from the sun.

Again, the energy for circularity of rings on planets had to exist and orginate from the parent planet itself. There is no other source. It cannot be supplied from an outside force. If enough influx material drifted into Saturn to form its rings the motion energy loss would be horrendous. It is unthinkable — there could be no ring system at all.

The tremendous weight gain would throw the planets into new positions, orbital.

Circularity holds the universe together. Everything is held together by circles. Most things are round. All forces are developed by circularity. The atom, which everything is made of is held together by circularity. Its internal forces are circular, and would not exist any other way. Rotation holds the solar system together. Did not mechanical locomotion begin with the invention of the wheel? We can always roll down hill, never up. Think about circles and round, now, and as the days pass and see how many things and actions there are without circularity. A combustion engine rotates as does a jet engine. A tractor or any mechanical device operates on wheels and bearings, and can only bring energy to a use by circles. A drilling rig cannot make a foot of hole with rotation, circles. Everything rotates. Planets are round, they are held in place by circularity. Space flights are not possible without circularity. Weather patterns develop as circles. Not much it seems is possible without circularity.

There is no such thing as a straight line in the universe except in short distance. Your home may be nicely squared up with straight lines, yet as a line lengthens it must curve. If circularity did not exist nothing else would either, perhaps not even the universe. We must have circles or oblivion. Does any force act without circularity? Circularity causes forces to relate to and affect each other for a result. Isn't tangibility a result of circularity? Ultimately, did anything exist before circularity? If things are only relative, they must be synonymous and contemporaneous, one cannot exist without the other. Trajectory and circularity must be the same thing. Neither can exist permanently in distortion. On the other hand, if all forces and actions could exist permanently in distortion, where would there ever be order and design to anything. We must apply all of this to circumplanetary development. Their beauty and orderliness were not brought about by imbalance and distortion. Annular theory is orderly and maintains relative balance during actions.

The idea of a tenth planet or some other object exerting a mystery force or tug on the solar system is reasonable. Orbiting irregularity indicate such. It is not possible to know all interactions and internal force of the solar system. Forces may differ.

So far, aren't we just talking about the old well established Einstein theory of "relativity?" If each planet gained new material independently from, and at different times than its neighbor, we would have disarray and distortion, not relativity and neutrality of force. This has not happened, we do not have continuous dissarray among the planets. There seems to be only one constant factor — motion —

do we need any other? Relativity and annular theory is our key.

The exhaustless and measureless energy of heat exerted to vaporize refractory minerals and metals, in the archean igneous earth, must have driven these vapors so far into space that, as they necessarily obey the mighty impetus given them by the orb's rotation, they accumulated so much energy that they must continue to revolve independently about the central body for a long time after the mass becomes cold. The day is not too far distant when it will be admitted on all hands that ring or annular formation is an indispensable part of planetary evolution. We see this legitimate energy beautifully exemplified in the clockwork of the skies, Jupiter and Saturn, twin giants of the solar system, proclaim this eternal truth. They are glorious in their procession in the heavens. Titanic worlds yet unfinished. Their future sedimetary crust is yet revolving about them. A thousand volumes cannot reveal it all. If geologists and astronomers will examine the record with an impartial view and philosophic mind they will see the beauty and simplicity of the creation process.

I cannot let the subject of circularity rest without making another point. The subject supports annular theory so well. If, as science claims, there is no such thing as a straight line in the universe as length is gained, it must curve (this seems reasonable). If objects, particles, substances, or things could forever move in a straight line from its point of origin homogeneously, or otherwise, or a noncurving azimuth, whatever that may be, just so long as our factor never had a curve which ultimately becomes a circle, would not the universe proliferate away from itself to the death of the universe?

On the other hand, this is not the case. Everything must move in a curve or circle which a curve becomes, or join a family of curves in a trajectory, and become circularity. May I suggest, that this curving circular factor establishes an imaginary invisible boundary around the universe, which could hold it together for all eternity? If, as some scientists claim, the universe is still expanding nothing will change in this hypothesis, it just has not reached its permanent boundary as yet. Why then, would the universe ever need to fall back upon itself for the second "bang"? If, velocity and gravital remained constant, and the awesome transformation of matter and energy continued, why couldn't the universe remain for eternity? I cannot prove this theory; however, no one else can prove their's either. If all forces reached a neutral area in time why must they do anything, or choose between two options.

Doesn't this again, bring us right back to Einstein's law of relativity? Everything is tied to relativity and gravital association, add if you wish densification. Imagine if you will, the imaginary boundary of the universe I have suggested, and ask why anything would ever move beyond that boundary, in that there is nothing out there for a gravital attraction to unite with. The question is, could anything escape from the universe? Where would it get the energy necessary to escape from the universe, when the universe is the only source of energy available? We are back to perpetual motion, quantum mechanics, thermodynamics, none of which will move uphill without an outside force. In universal dynamics where would a force come from greater than what the universe could supply?

If we are to suggest enough energy could be concentrated in an object to thrust it beyond the boundary of the universe and the energy must come from the same source, are we not suggesting a flaw in universal law whereby the universe could disintegrate? However, the reverse is true, and any concentration of energy must be toward the center of the total mass, not away from. Looking at the reverse again, if the stuff of the universe with less energy moved toward the boundary it could not escape because of lack of energy to break from gravital law to an area with absolutely no gravital force. Isn't it possible then, nothing could leave the boundary of the universe? The unthinkable is that the universe is boundless and goes on forever into space. Then what difference would this make? As Einstein would say: *"God does not play dice."*

The arduous fascination in search of the ultimate particle has led us into many no man's land areas. At this juncture I would say man may never find the ultimate particle. If more proof is found I will change my mind on this. Magnetism has eluded man since hundreds of years B.C. What is the basic particle of magnetism is the long sought-after answer. The neutrino and magnetic momopole come into play at this time in history. The magnetic monopole is thought to be a remnant of the Big Bang. If the Big Bang is a true concept all things and factors came from it. Magnetism is still an elusive factor. If the subatomic particles such as the neutrino and monopole are proven, will this answer or create questions? Does the neutrino and monopole contain mass is the question. Superconductivity and electromagnetic phenomena exists in spite of the knowledge of quantum mechanics and other theory. (Quantum, is everything quantized?)

Besides the Big Bang we now have "The Big-Bubble Theory". Astrophysicists and cosmologists who accept the Big Bang theory must agree the universe is homogeneous in total concept, and must remain as such, as how could the universe fall back into itself for the second bang as some claim? Could the universe be shaped like an ice cream cone or a shoe? If it is, it could not fall back in on itself as some claim it will? (I do not.) An explosion in midspace must have homogeneity and remain as such in totality. The law of acceleration will carry the mass to its boundary. The law of convergence will bring it back to a common center. Of all the theories the Big Bang would be the least likely to violate these simple laws. The earth is heterogeneous, just as other planets and satellites are, yet the universe in total concept must be homogeneous, if the Big Bang theory is true.

After a consideration of all of these theories, when we apply gravity, centrifugal, acceleration, and convergence, aren't we right back to Einstein's "relativity"? If we do not accept homogeneity, we must abandon the Big Bang theory.

CHAPTER 13

SUPERNOVA HYPOTHESIS

Collision destroys energy, never enhances energy. A very simple example is an automobile accident. Planetary systems could not be built by collision.

Everything that happened in earth geologic development and dynamics must have happened twice. The igneous molten earth in development process cast off material which formed rings and belts around itself in an orderly system. Great heat and centrifugal force would segregate the material into specific rings and they would return to the earth in the same order as they arose. Crustal metamorphism would be caused by heat before, not after it returned from space rings. Some blending, however, would be possible. Crustal upheaval would cause geological conglomeration, as well as radiation inconsistency. Did radiation variation develop (such as iridium or other) with the stratum or after?

Annular theory is so reasonable and answers so much. Many scientists come very close to this theory and stumble over it. An article on the subject suggests that the mass destruction of marine microfossils and many reptiles may have been caused by billions of tons of extraterrestrial material suddenly falling onto the earth. Here they stumble over the annular theory as the question they ask, where did the material come from? What was the mechanism of its arrival on the earth they ponder? Did the flux of material come from within the solar system, or from outside the solar system? They admit there is very little data available. The elaboration of this hyposthesis is complicated, yet they procede anyway. It is suggested this material may have come from a gigantic stellar supernova explosion from outside of the solar system. Would such a gigantic explosion produce only micrometeoric material and how could it be transferred to the earth in such a large amount? Why would material from interstellar supernova explosion accumulate in sufficient dense concentration as to arrive at the earth? Could such an accumulation of material meander through the solar system until it struck the earth? In this case, it would not feed its material to all sides of the earth. Enhancement of the material to the degree it could not miss the earth would fill most of the solar system. This hypothesis fits the annular theory. The material in question arose from the earth and later returned to the earth. To destroy a hypothesis or theory we must examine all possibilities, here we have only two. If we are to say this material was highly densified into one bunch it could not cause the desired results on earth. If we take the proliferation of the material it cannot accomplish the known results which now exist. We must abandon the entire theory. If we accept the theory that outside influx material developed circumplanetary ring systems, there will be both borrowing and cancelling forces, which will exert negative resistances, potential inherent forces cannot overcome. To energize a secondary system from the total energy volume will weaken the vigor of the energy unit, not enhance, which would be required in this theory.

Is the crust of the earth and the atmosphere only distillates of the mantle? We will answer yes, only if we apply it to the annular theory. How else could it be as it is today? Carbons were distilled in the igneous earth and thrown into rings around it over a very long time span and returned the same way, giving the earth many different oils and coals and gases as deep as six miles. To say these gaseous compounds from the mantle formed into organic material is not possible. Organic compounds and life material are separate from inorganic to this very day. Organic molecules need photosynthetic processes. Inorganic material will not evolve into organic by itself. Fossils in carbon beds are foreign material. We cannot fit the supernova theory or plate tectonics into this picture.

The density of the mantle increases with depth as the annular theory will support. It is reasonable to say density increases to the center of the earth. Scientific thought says the density does not increase smoothly, however, the curve of density displays distinct steps. Variation in the distribution of mantle material as well as crustal material to surface must be true as applied to annular theory. Ring systems returning to the earth would cause this separation factor at development, not after.

In reading an article, again, I see how close science comes to the annular theory, I just do not see why they do not accept it and take things from there. Everything will fall right into place. One authority says in discussing plate boundaries including divergent and convergent types below the crest of ocean ridges, etc.; *"At convergent boundaries plates may collide, pushing up crust to form mountains, or one plate may move under another, carrying lithospheric (rock) material back into the mantle."* I would apply this idea to hydrostatic compression due to ocean augmentation, and other material as well, from ring declination. Where else could the material come from? This replaces the supernova theory as well. Again, where would we get an energy force for horizontal lateral plate movements? Everyone knows gravity on earth is not a horizontal lateral force, unless we apply a 90 degree lever action to it. In plate tectonics where do we get the lever action? There is no such action. In physical law we must always account for the direction in which a force is acting. Verticle actions are easy to account for, as everything that goes up must come down. This is just simple gravity. Horizontal forces too, must be accounted for, and they are not so easy to come by. To force lithospheric (rock) under continents or for that matter, ocean floors, requires a stupendous force, not accounted for in plate tectonics. The only known force great enough to do this task is hydrostatic compression equal on all sides of the continents, caused by ocean augmentation from annular ring systems around the earth. The strongest position the theory of plate tectonics has is propaganda. When known physical law is applied to the idea, it just will not stand.

The experts are right in claiming the earth is very dynamic, I agree. The rotating sphere has physical properties which indicate its great mass concentrates toward the center. The very powerful gravitational field of the earth dynamics has to come from a great concentration of mass towards its center. In looking at a cross section of the configuration of the earth as depicted by one publication, based on plate tectonics, some questions arise. First, do not let big words throw you, they are just another way of saying simple things. We must consider the direction of the action of the force of convergence. Can its direction vary? Yes it may. The diagram of the earth's configuration I have before me shows the lighosphere (rock surface) moving both directions from the Mid-Atlantic Ridge, with the asthenosphere (hot material next layer below) doing the same thing. As even a child would know, if we have two layers of material here moving away from each other without displacement material, why should there be a Mid-Atlantic Ridge? In this case it should be an increasing syncline as long as the material on both sides kept moving away. To move material away from an area will not create a ridge or even an anticline, will it? You decide.

My mind has never been at ease with the theory that billions of billions and trillions of tons of the earth's plate and mantle material is constantly moving around the earth. There would have to be a jam-up somewhere, perhaps on the other side of the earth. Why has the earth remained spherical all of this time with the exception of a few mountains which will about average out gravital, axial? In annular theory, material leaving the earth at the equatorial plane, forming belts, rings and disks of material, then returning to the earth at the centrifugal equatorial plane, would keep the earth spherical. The supernova theory has been interesting, yet it falls far short of fulfillment of this great dynamic requirement. Our agent must always be equal to the need, not too puny.

The stratigraphic column has at least 12 major geologic ages from Cambrian to Tertiary, with dozens of identifiable interrelated ages within the geologic crust of the earth. If the origins of the continents is merely an upthrust of asthenosphere material, when did these geological beds, Cambrian to Tertiary develop with their minerals? They had to be stratified independently. Plate tectonics or supernova could not accomplish this magnificent structure of variations. In research a vast amount of written material on continental drift, I have learned very much and proven nothing supporting the theory. In a few words, the ideas are a total contradiction of the laws of displacement and convergence. No one need be a university graduate to arrive at this conclusion.

CHAPTER 14
CONTINENTAL DRIFT OR UPHEAVAL

Do you know of any lateral energy force great enough to pull a continent loose from its moorings, so that it may drift aimlessly in the oceans of the world, while hydrostatic compression from the same oceans held it locked to a permanent position?

There are some who claim the earth's crust rests on a hot plastic, rubbery underlayment. This is given as a reason for the claim that continents are continually rising and falling; alternating from mountain peak to ocean bottom. This plastic material they claim does not have dimensional stability. Can such a theory be correct? Let's ask a question to answer a question. Let us say the deeper earth is rubbery as they claim and too hot to solidify or crystallize. Have they discarded centrifugal force known to exist by all persons? If thousands of miles of ocean water, miles deep, covering 71 percent of the earth, will obey centrifugal law, with only a balancing tide each day, why should not this rubbery material, supported on top and bottom, equally distributed around the earth, do the same? If less stable material as water will remain in a circular position around the globe, in obedience to gravital and centrifugal law, where do they have grounds for lack of stability and equilibrium in this other material better supported than water at the surface? Where does a difference lie, except against their theory? Are they to argue that a cubic foot of water and this plastic material may have a small weight difference would cause severe abberations, mountains may alternate from 30,000 feet above sea level to ocean bottoms 30,000 feet deep? We are talking about 60,000 feet of differential of earth, water and rock, 30,000 feet either side of datum plane by elevations. If there is a small weight difference in this mantle material from rock, water and other material, isn't it subject to gravital and centrifugal law? They must claim this plastic material is fairly evenly distributed around the globe. Then what reason would there be for it to violate gravital and centrifugal law, any more than water will? Some very large, clear span buildings are constructed in an arch so their own weight holds them as such. Next time you are near a pool table, take a cueball and try to squeeze it in your hand to a smaller size.

Some persons say the world is now in the same unstable situation today. They say the crust is again out of a state of equilibrium. The continents are again ready for buckling. The trigger is a matter of time, they say. There is that old, overworked word "time" as an excuse and reason again. Annular theory will answer these questions and do away with such unreasonable theory.

The very strongest argument against continental drift theory is that it is an absolutely unprovable concept. There is no working model anywhere on earth is there? This would be true of many scientific ideas nowadays. The annular theory has a perfect working model in Saturn and other ringed planets. Often the most difficult problem may be solved by a very simple illustration. Let's compare continental drift to such an act as loading a large refrigerator into a small pickup truck. Let us say a twenty foot long pickup truck is backed up to a large refrigerator seven feet tall. Two strong men lift on the far end of the refrigerator to bring it up to slide it into the pickup. The driver, we find, forgot to place the shift in park position and set the brake. Instead of the refrigerator sliding into the pickup, the truck rolls forward and the refrigerator falls to the ground, as the pickup stops rolling from the lack of force. We now have a mass 27 feet long because our mass drifted.

Repeating the process with the driver setting the brake, they are able to lift the refrigerator up into the pickup. Our mass did not drift so we now have the same mass, though higher. Comparing this action to continental drift, it would be the same. If the great force that caused the western overthrust belt had moved the continent, or if any other overthrust on earth had done so, how could we have an overthrust as some claim up to four miles thick or high? If the continents moved from side pressure, we would have a longer, lower mass, not higher, shorter as now actually is the case. Because the continents have not moved from side pressure of ocean augmentation all around them, they remained as an arch, and were continually thrust higher, never lower. We now have great mountains to fulfill the law of displacement in the crushing force of annular rings dashing into the earth, in equal proportions as a spinning wheel suggests. In the proven annular theory, wherein material from ring systems in rotation on the earth at the equatorial plane supplied material to the earth also in rotation, we have an equality of development. Water will not violate axial law, but would rather seek it for worldwide continguity. Side pressure would be so nearly equal on all continents, it would not be possible for them to drift. Just as continents today have ocean water contiguity, this would have been so in geologic times. Oceans around the continents give them stability as to drifting, yet may be the cause of great upheaval of them.

There is an endless amount of scientific information available on continental drift and plate tectonics. Because most of the revolutionary science of the last decade supports my accepted theory of annular development, I will only answer some of the points postulated. Please read all of the details for yourself in these other publications. Scientific American has a lot on this subject and others. It is very interesting material, though most of it will support annular geology, not continental drift.

Formerly, most scientists thought the earth to be a rigid body with fixed continents and permanent ocean basins. Many scientists regarded continental drift theory as heresay up until about the 1960's, according to Scientific American. After a meeting of geophysicists held in Moscow in August 1971, it was found many scientists now accept the new theory of continental drift — though it is not new. My theory of ocean augmentation will answer all of the puzzling questions not now answered. Continental drift does not answer them. It is plain ocean floors have been spread apart, in which one plate overrides another, forming trenches and islands and mountains on the ocean floors. The crumbling of the surface forms mountains and plateaus, they say. This has been my stand all along, the force of water-ice and ocean augmentation caused this. Yes, this is a revolution in science. Accumulated scientific observation will show this to be superior to the accepted theory. The abandonment of Ptolemaic theory was not easy.

Copernican theory replaced Ptolemaic theory which opened the gates for Galileo, Kepler, Newton and many others. So it is with my accepted theory of Annular geology. Once the qualified scientists accept this theory, they must help fill in all of the details. They are well trained for this.

ISOSTASY (HYDROSTATIC COMPRESSION)

For example, spreading sea floors with lateral plate movement is not continental drift. The same force would be exerted on all sides of continents by ocean augmentation forcing material laterally under the continents for continent-building actions, upheaval. Continents did not break up — they just got smaller in size and much higher. Continental fragmentation is only a theory and not provable. Sea floor spreading is a necessity with ocean augmentation and continental upheaval. Geophysicists must admit the earth's crust has been folded, buckled, upheaved, with shorelines always moving up. Continental drift fails to account for this action, annular theory does. We know what we see. If the earth is not as rigid as once thought, what will change? Scientists are mentioning separations of continents of 1½ to 3 inches at most per year, and may I suggest, this may not be static or could even be reversible. The top of a very tall building may move a little one direction and five years later it may move the other direction. The law of displacement is our key. Continents do not drift around and collide with each other as icebergs in water. Compression does effectively bouy up mountains as we shall see.

The theory of isostasy is still sound, with a few variations at this time in geologic history. The oceans and continents do float in approximate hydrostatic balance with each other. As the oceans were augmented with more and more water over the millions of years of development, this fact would be more and more applicable. Oceans do have contiguity as anyone would know. Gravital and centrifugal force is balanced. Webster's New World Dictionary says about "isostasy"; *"Approximate equilibrium, in large, equal areas of the earth's crust, preserved by the action of gravity upon the different substances in the crust in proportion to their densities. A condition in which there is equal pressure on every side."* This describes perfectly the situation we have on earth today. A view suggested by Isaac Newton is very interesting. He felt the cooling process of the earth created compressive forces, which at intervals, squeezed up mountains along the weak lines of the continents. It was believed the cooling earth would shrink some to start this process, and perhaps it did. Isaac Newton's theory fits perfectly the annular idea of earth development. The continents have been squeezed up by isostasy law.

The hundreds of theories still being postulated and borne along, do so because the majority of scholars and scientists do not recognize a basic fundamental truth, annular geologic development of the earth. Was not this the problem with Ptolemaic theory? It requires such a crushing amount of evidence to unseat a long accepted idea, even though it may be wrong from inception. No matter how much factual information is later heaped on a wrong premise, it will never fit. Because of a wrong premise a simple fact is a major stumbling block to learned men. What about submarine geology? Because such are found far inland or on mountain peaks they feel the mountain peak was once an ocean bed. The last great ring of water fell in on a tropical world and floated this life inland as the mountains were thrust much higher. Many biologists have spent much time studying land bridges between continents in an effort to account for distribution of life around the earth. When ocean levels were much lower, a map of the earth would appear much different. They assume as ragged edges of two pieces of paper torn apart may be fitted together, so it was with the continental drift theory. Meteorologist Alfred Wegener, in 1912, argued there was a different primeval arrangement of land mass, with all of the continents joined in one supercontinent. My mind rebels to this idea instantly. I will not ask him how the supercontinent got there in the first place. My first question will be, where did the force and source of energy come from to break this supercontinent up in all of the lands and islands of the earth? There is no accountable reason available to a sound mind, is there? If all of the land mass was in one continent and if all of the water we now have in the oceans was here, would not the law of isostasy be a hundred fold more persistent against this single continent and keep it forever trapped where it found itself to be? I hope nobody will be so foolish as to say the master continent just drifted aimlessly, like an iceberg and slowly broke up into pieces and left them trailing around behind her about the globe. We all may stretch our imagination at times, however, this is lunacy.

GEOSYNCLINE — ANTICLINE — SYNCLINE

There is some very interesting material on Alfred Wegener in Scientific American; "Continents Adrift and Continents Aground." Please read it for yourself. Everything he says will drive you to the annular theory. Another question I must ask; is where were the great carbon beds of coal, oil, gas, located while he has continents breaking up and drifting all over the earth? Did they form before or after, and do not forget some are as deep as six miles? There has been a struggle between paleontologists who rely on vanishing land bridges, and geophysicists who insist the sinking of such bridges was impossible.

This reminds me of the evolutionists and creationists who are in court to find out who is right. They are both wrong as are the two I mention. One will claim continents are forever rising and falling, becoming ocean bottom and mountain peak. Others say a continental bridge cannot sink. Which is right? Again, both are wrong. That is why the problem is not resolved. A very simple answer is that of ocean augmentation, which could easily cover a land bridge. Today, even though the motor that moves continents is not known, anyone who rejects plate tectonics is considered a reactionary. Wegener said: *"I do not believe that the old ideas have more than a decade to live."* I will say the same; I do not believe vegetarian geology will live another decade. I am an amateur, just as Einstein was, and many others. Everyone was born an amateur. An amateur is anyone who does not go along with accepted theory. An expert is anyone who will accept a false premise and build around its perimeter for the sake of acceptance and a job.

Because of the apparent jigsaw-puzzle fit of the coastlines of Europe, Africa and the Americas, it is almost universally accepted by geologists that they were once joined together. My comment in a constructive way is, just because so many accept a theory does this make it a fact? In my lifetime, I have taken the opposite view, as the number who believe something does not add to its validity. Often it just means more propaganda has been given to this viewpoint. Many times we have heard of someone finding a stain on a board or piece of cloth or any object and many believe it to be a valuable relic of religious veneration. The truth of its origin and value may never be established, yet it may go down in history as a fact. We in science must not be guilty of such acts. The shape of the continents on each side of the Atlantic Ocean would be the very last proof I would accept for continental drift. Others say the Great Flood caused continental drift. Most scientists dismiss this because they feel it is a supernatural phenomenon, though they have no proof either. There has been a great flood of water on the earth; however, this would not cause continental drift. This would cause continental upheaval. It would, however, increase isostasy factor for continental stability. I would never dismiss continental drift contemptuously or any other theory for that matter. I would present facts and reasons for not accepting an idea as a basis. If Wegener had had a chance to study and look at annular theory (rings on planets and the earth) I feel he would have accepted it. If the truth of an assertion may only be tested against a prevailing set of rules, then we are in a heap of trouble. Sad to say, oftentimes there is an immediate pressure for an answer in issues of science. It may be economic or competitive pressure, or the pressure is the reason for acceptance in that there is not a better answer at the moment. The fossil find named "Lucy" is just such a case. We must outdo the others now — even if we invent proof. Some may feel they will reject a new theory which may undermine their own, in the hope it may be proven wrong later.

I have gone through mountains of information and data on these subjects, it all points me in one direction, the annular theory of Isaac Vail. Affinities of fossils and rocks do not support land migrations. It is a hokum as is speciation. Endless diagrams, figures, and equations are presented, all of which is only theory. These are applied to geosynclines as proof for theory. Are synclines and anticlines the origins of mountains as is claimed? May we reason thus. Are the continents continually rising and falling and drifting? It is a dictum of geology that geosynclines eventually become mountains by folding. This is supposed to be a very slow process over millions of years. Therefore, this cannot be correct, as mountains are built by a piercement action, in a short time span cataclysmically. If continents and mountains are slowly rising and falling over millions of years, why are there no unfolding geosynclines on earth? There is a very productive oil and gas anticline in the great Williston Basin of North Dakota and other area states. When was the oil placed in the basin at many levels and was it ever a syncline? It seems a geosyncline agitated by so-called plate tectonics cannot be a mountain building cycle. When did synclines trade place with anticlines if this is the case? If this syncline cycle is true, what would happen to our oil fields? Where would we find them? Why do geologists look for oil under a seismic high if geology is always trading places? Geosyncline collapse just cannot be the prime mover for mountain building. Looking at a mountain we can see it has been heaved up piercement in structure. The broken upended rocks at such elevation are not a slow erosion cycle of geosynclines. Can you imagine the great western overthrust belt being built by such a slow, weak, puny agent?

The greatest mistake of the geophysicists is they do not account for the energy and inertia required (usually lateral horizontal) for the actions developed in their theories. The first thing I want to know, what is your energy source? If it cannot be accounted for, the theory is worthless until it is. Gravity is not a lateral force, unless you may be skiing down a mountain. Continental drift must always act from datum plane or below. Gravity may cause a lateral action. Instead of putting six inches of water in your bathtub, fill it full and gravity will cause the lateral pressure against its sides to increase. This was true when the oceans were augmented with additional water. This increased lateral pressure against the continents on all sides, squeezing them up higher. There is hydrostatic balance around all of the continents, just as when you get into your bathtub. Lateral pressure will increase on all sides of the tub. Simple, isn't it?

In reference to gravity I here apply it to earth only. If you parked your car on a level place, it will not roll away because of gravity. I do not mean there is no horizontal force at all. If another car were parked in front of yours, there would be a small gravital attraction, though the cars would not roll together. Two large boulders placed near each other would have a small gravital attraction, still they will not roll together.

The flow of a river is caused by gravital action and force. The rivers of the world may have, at times, only a small regional drop, yet gravity will cause this near horizontal movement of water. The action of gravity is fixed on the earth. When we refer to a gentle drop of a river and steeper in other places, it must always be to lower, not higher, unless, however, there may be an area with a canyon wall to hold the water until a lower elevation is approached. We must account for the curvature of the earth in our equation. In a small area, water will always be level. In building structures such as the pyramids, a trench was built around the base and water put in it to establish a level foundation. A plumb bob was then used for vertical straight up and down. When a river is dammed and backs up for 100 miles, we always say the water in the reservoir is level. No, it would have to obey the laws of the earth's curvature for a 100 mile distance. All things are relative. As the Mississippi flows from its origin in Itasca Park in Minnesota to the Gulf of Mexico, its drop must always be lower than the curvature of the earth. This is not a great revelation, but interesting.

There is positive and dramatic proof that the continents have been upheaved and even upended by great cataclysmic forces in geologic times, of which only the annular theory will account for. May I draw your attention to an ad in the September 1981 issue of Newsweek Magazine, by Tenneco. They say in part: *"Nature lifted, folded and fractured the underground rock strata; potential energy-bearing formations are tilted almost vertical. To find oil we cannot drill straight down as we normally would, because we might drill parallel to an oil-bearing formation. So we start drilling vertically, then force the drill bit sideways at a 40 degree angle, hoping to intersect the oil bearing strata."* This is very interesting, and they have found oil this way. What forces upended these mountains and the oil field in them? It was not done by simple long term erosion as claimed by some scientists. Erosion would destroy such a formation as it will only build flat lands. This verticle upheaval must have a cause competent for the task.

CHAPTER 15
ANAEROBIC BIOLOGY

In evolutionary science, there seems to be a desperation to prove how non-living matter became living (organic). Vividness of argument is obvious. We do not need great eloquence of words and speech when dealing with facts or truth. Facts are abrupt and do not need a lot of argument. They are powerful in their own right. Science in its attempt to build a bridge from non-living matter to living, without a vehicle or motor must use thousands of words — they are building a word bridge, because they have no other material to work with. Organic material stands apart from non-living matter. Living organic life, such as a tree, can draw sustenance from non-living matter because it has a photosynthetic motor. Never may the process be reversed, even via thermodynamics. Science is just moving in the wrong direction.

Some scientists say solar energy was sufficient to cause life to develop. Yet, they do not explain how sunlight could have been used for molecular synthesis without a metabolic motor. Just as the energy of gasoline cannot be used without a motor, neither can sunlight be used by non-living matter because non-living matter lacks a photosynthetic motor. If a contradiction is found in the second law of thermodynamics, it lies not with life, but rather the evolutionary model for life. The creation model stands, where the evolutionary model does not.

We are hearing a lot about anaerobic life nowadays with the research from ocean bottoms and Antarctic regions. Some feel a life has been discovered which is independent of known life processes. Is this really true? Anaerobic life must get oxygen by some process other than normal free oxygen. Anaerobes could get oxygen by the decomposition of compounds containing it or stored away in times past. Life forms do not live outside of, or violate the boundaries of life. Some appear to, yet do they? Life only comes from life. Some animals living deep in the oceans seem to violate the boundary of plant life, yet they do not. The great and fantastic process of earth development must have placed many usable elements in the earth for life to use. This could be a latent process the lower forms of life may draw on. Some very courageous and smart scientists are doing research in the Antarctic. They are finding algae in rocks and lake bottoms. To my pleasure, everything found so far, is just as it should be, in my accepted theory of annular geology. (Smithsonian, November 1981). The question is not how this could be, but rather how could it be otherwise? They saw certain organisms growing only under the north side of rocks — the side that gets direct sunlight. Doesn't this answer most of our questions? Nothing here seems to be out of order. These organisms live in many countries around the earth. More research is needed and it will prove a single source of life, not several.

Much more could be said on this subject, though I must comment on the aftermath of the explosion of Mount St. Helens, and the findings of microbiologists there. There is a prodigious explosion of algal and bacterial life in the wake of the eruption. In annular geology, this is how it should be. Not sterile as some felt it would be. Life did not live in this hot material below ground. (National Geographic, December 1981.)

In this volcanic material we have the ingredients for life to develop. Add warmth, sunlight, and oxygen, what else could result? It is compared to the life found in the hydrothermal vents deep in the Pacific Ocean in the East Pacific Rise. The mystery of such life found in the Antarctic, deep oceans, and now at Mount St. Helens, is not yet fully answered. Is there a connection? Do they live from the same source? Off hand I would say yes, and stand to be corrected if later proven wrong. Do we have another source of life and energy but solar, even if latent? There has been a latent process built in we do not as yet understand or connect. Much has been said about fossil life in the Precambrian rocks. There are many unanswered questions. When did fossil life, if it did exist in Cambrian rocks, get there? Yes, how did it get there and when? Estimates of three to over four bilion years is a long time. If pressure and heat were as great as claimed, why were the fossils not destroyed, or metamorphised, if I may borrow the term? Nature does not perform the way we think it should; we lack a key to unlock it. What was the sustenance of these small fossils and how did they get there is a fascinating question. If we ever find the answer it will be simple as usual.

VOLCANOES AND CAUSES

Is the volcano the shaper and sustainer of the earth? Yes, volcanoes helped in earth building, but are only a minor factor compared to the great energy and force used and needed. Volcanoes are only the tail end of a much grander and larger process, called Annular Systems. It is true, erosion would reduce everything to level plains. In converse, are volcanism and earthquakes the uplifter of the land and father of the mountains? Volcanoes have added some material to elevations in a few areas, yet neither volcanoes nor earthquakes are competent for the monumental task to upheave and build mountains of solid rock thousands of feet high, and raise continents into existence. It has been suggested that fine volcanic ash high in the atmosphere can regulate solar heat and energy reaching the earth. It would certainly have some affect on climate, but could volcanism be the thermostate of the earth? Perhaps not, though pollution is helping in a strong way to destroy climatical balance. It is estimated 200,000 lives have been lost in the last 500 years to volcanic actions as a measurement to show its great force. Perhaps millions of lives have been lost to earthquakes and they have added little to the upheaval of the continents. The great energy force needed to shape and build the earth originated in the igneous hot earth, with some of this energy, latent though it may be, spending itself now in volcanoes and earthquakes. Compare in a small way, Mount St. Helens to an entire hot, molten earth throwing material into rings around itself, later to descend on the planet, as the ash of Mount St. Helens did. This, then, is a competent world building process.

Great errors are sometimes made in the equational process used when our model is too puny for the job assigned. The only alternative is to use a longer time element which may very well destroy the theory. Who can argue the difference between three or four billion years. I do not mean to minimize the power and destruction caused by volcanoes. In Iceland, where the Mid-Atlantic Ridge comes ashore, there have been many volcanoes. New ones began again about 1975. The earth must contain a lot of magma from the fires of its early origin. Geothermal energy may break forth and send volcanic lava to surface at times. There will be some surface oscillations as a result. If one small area drops, another will rise. Walk across your waterbed and you will see the same thing happen. We cannot say the Mid-Atlantic Ridge caused North America and South America, to separate from all of Europe and Africa by thousand of miles as continental drift. Why do such never take into account the other side of these drifting countries? After all, Alaska is still connected to Asia. Perhaps Alaska kept pushing Asia along. But, Asia is connected to Europe, and Europe was supposed to be separating from the Americas. This brings us all the way around the world, doesn't it. Who is to say what was separating from what, is the unanswered question.

Some outstanding volcanoes in history are, Vesuvius in A.D. 79. A fifth of the people and animals on Iceland were killed in 1783 by a volcano. Krakatoa was the greatest eruption in history in 1883. It killed 36,000 persons. Not all is known about this subject and may well never be. As energy force of volcanoes and earthquakes spend themselves, these actions will also diminish.

SOME ENERGY DYNAMICS

Embodied in the many ideas for energy are theories as yet unworkable, and a pipe dream, though they have possibilities. Ocean

Thermal Energy Conversion is one (temperature difference in the ocean from surface to 3,000 feet deep). The oceans of the world have their own laws, though not removed from annular law and other fundamental law. It has been said: The perpetual bond between sea and air is nowhere more intense than in a water spout. It is as if, for a moment, the creation of the oceans by condensation from the early atmosphere could be reversed and the two great sustainers of life be made one again. This thought had valid application long ago when earthly ring systems were forming from the igneous earth; however, today we do not have an energy force which will throw earthly water into rings around the globe, nor does evaporation have enough solar energy to do this mighty task. We may be thankful it will cause weather patterns for rainfall.

There is the great energy and power stored in the movements of the oceans of the world. To understand the basic interconnection of natural phenomena, and understanding of its origin, which is annular. It has been my privilege on several occasions to spend time at the Princes Hotel in Acapulco, Mexico. This hotel is not located in the harbor, but on the open ocean, down the coast a ways. This gives one a perfect view of the ocean and the ships traveling each day to and from the Panama Canal. Sometimes we would spend a few hours lying on the beach under a shade, where I would study the constant motion of the ocean breakers. Worldwide, this would be a lot of energy, if it could be harnessed. This would be true of the ocean currents and other movements of the oceans. Again, we need a motor just as gasoline does, or photosynthesis. This energy may be used and is now being done in a small way. In some mass scale projects, our mechanism to harness may be so monsterous in size, we may spend nearly as much energy developing the project as it will produce. Here we must consider the first and second law of thermodynamics. Are we robbing Peter to pay Paul and holding out a little in the transaction? Is the "Unified Earth Theory" the answer? If there is nothing to gain on either side of an action, why do anything at all? The energy to build our machine and the friction to run it, may cause a break even or loss in total concept. We must have a continual flow of energy, which we do have in the movements of the oceans, if it is not all wasted in functional processes through the entire chain of use and delivery where needed. Does our action have possible use or are we just trading nickels? Many man-made projects today do worse than trading nickels, there is a loss in totality. Many dams we find this to be true. Nature in its own setting will often outproduce them. The farther forward we carry this premise, the better it becomes for natural processes, as all dams will eventually silt shut. The larger the dam, the greater the silt factor. The second law of thermodynamic tells us energy transformation leads to nonrestorable reduction in its capacity to do work. The attribute of energy to do work is not recyclable. We need a continual flow of energy. We do have a continual flow of energy in rainfall over a twenty year period. A dam trades perpetuity, a short term gain, for greater long term loss. The tragedy of engineers is, they lack knowledge of true physical law. Just because something can be done does not mean it should be done. They are spending other people's money. Economics do not count to them. More on this subject later.

Many are those who have experimented with perpetual motion as an energy source. We must rely on solar for this. Consider the "biomass pyramid." Plants always must produce more weight than the herbivors who live off them (man or animal). Plants grow faster than the animals that live off them, they do not reproduce as fast. After all is said and done, the universe is a great source of energy — a great ecosystem we all live from. We are tied to the earth's ecosystem, if we destroy it, we perish.

An article in the December 1981 Smithsonian says: *"Nothing may turn out to be the key to the universe. There's more to a vacuum than meets the eye. The vacuum may hold the answer."* The idea is the universe and all energy forces may be rolling downhill as on a flight of stairs, to a lower energy state. That, after the big bang, the universe is here for the same reason that a ball rolls downhill. A vacuum, it suggests, may not be a vacuum at all. Which is, more or less, unthinkable, something or nothing? We could argue for an eternity which way is up or down. We will run out of positions and directions before we find the answer. Science is now speculating that a vacuum (nothing) is something. At least this line of reasoning will stimulate thought in my reader as well as myself. You may know as many answers to this as anyone else. Evolution and science deny the existence of God. Now they say a vacuum (nothing) is something. Perhaps then, in their last discovery, they have found a place where God and spirit creatures could very well exist. Science and technology claim they are the highest authority in knowledge and wisdom. Perhaps they have undermined their own position in this last discovery.

SPECIATION AND GENETICS

Adaptation is not speciation. Adaptation remains within the genetic boundary. Science says most speciation takes tens of thousands of years or even millions of years. Small wonder this mystery of mysteries has not yet been documented in the more than one hundred years since Charles Darwin got biologists to thinking along these lines. This is a safe premise. One claim is made that climatical change caused speciation by adaptation. Then the claim is made that climates have been constantly changing which causes the speciation. If it takes millions of years for speciation when did adaptation ever catch up? Extinction of the species we are told is evolution. A quote by Ken Brower, from the book "Extinction" by Paul Ehrlich: *"When the vultures watching your civilization begin dropping dead . . . it is time to pause and wonder."* Another quote from the same book, *"We have not inherited the earth from our parents, we have borrowed it from our children."* Evolutionists are facing a stone wall; no new forms of life are now coming into existence by speciation.

In the modern day eulogy of technology, we are led to believe there are no misconceptions or scientific errors, the past is behind us now. The blunders of Aristotle and others will not be repeated. There will only be new discoveries ahead of us and these will always be right because they are so smart now. Is this line of reasoning really true? Are we really removed from the errors of the past? It would seem not. Look at medicine, drugs, and chemistry (Swine Flu vaccine, etc.). Recent TV programs such as the Donahue Show have presented evidence by experts which completely overturn accepted medical practice and drugs, showing them to be worthless and even dangerous to health and life.

We are told all known genetic knowledge is at our finger tips. Science now has the answer to most everything in genetics. Not so! A 79 year old scientist, Barbara McClintock has revolutionized the field of genetics. She has a Ph.D. from Cornell University. She works in a small building at Cold Spring Harbor, N.Y. as she does not like big labs and offices, according to an article in November 30, 1981 issue of Newsweek. The article goes on to show she used Indian corn to unlock the secrets of genetics. Having grown up in the rurals, we children used to pick this kind of corn and wonder why some kernels on the same cob would be red, blue, and yellow. This observation helped her solve the riddle of genetics. She pulled the rug from under some old assumptions. Geneticists assumed that the genes that dictate physical traits and functions of cells were arrayed on the chromosomes like pearls on a string. She concluded genes shift position from generation to generation. This is how the color of the next generation of kernels could be different from the parent. Cells would contain all of the genetic material that came together at fertilization, yet kernels could shift from red to yellow, and not violate genetic law. This then, would account for the wide range of and variety of life that exists, without violating the specie boundary. Those scientists endeavoring to prove new species by speciation, will jump on this as proof there is speciation in evolution. This attempt will prove to be futile as there never has been proven a speciation, kind to different kind in the history of genetic research.

Life kinds just do not develop or upgrade by chance mutations. It is not my intention to take any credit from this intelligent woman scientist; however, her findings are just as I had believed them to be for many years. The beautiful magnificent creation is locked to an unbreakable law for which we may be thankful. To tamper with this will not result in good. Recombinant-DNA technology will use this information on cells as a means to further their schemes, which in the long term will be to the detriment and sorrow of us all. We can all remember when they told us atomic research would be safe and under control. Ask the folks at St. George, Utah.

Most scientists and paleontologists believe and teach the theory of speciation. Yet they are not able to link life from one stratum to the next via speciation. They feel extinction of life is a forever revolving fact, part and parcel to evolution. New life they say, always replaces those which become extinct. Man's turn should come then. They say man is destroying life kinds faster now than the evolution process can bring new forms of life into existence. We know many life kinds of plant and animal have become extinct in recent times. When have you noticed a new life or specie come into existence? It seems clear that the annular cycle of earth development from Precambrian to Tertiary or surface has run its course and any life extinctions now is final. There is no life replacement. We are on the "rivet Popping Space Ship Earth", Dr. Paul Ehrlich talks about in his book, Extinction. Rivets represent life kinds. If enough rivets are popped, the wings will come off and we will crash. Dr. Ehrlich says: *"The end may come so gradually that the hour of its arrival may not be recognizable, but the familiar world of today will disappear within the life span of many people now alive."*

Note: Precambrian is 16,000 feet deep in the Williston Basin. Geologists say it is possible to surface small fossils from this depth. Unofficial report is a modest amount of oil has been recovered from Cambrian at about 12,000 feet deep in a Stark County, North Dakota well. The first oil recovery from Cambrian Deadwood sand in the Williston Basin. Yet, these scant fossils did not make the oil.

CHAPTER 16

DINOSAURS — GEOGRAPHIC SPECIATION

Yes, it is true, the fossil record is a testimonial to dinosaur extinction. They did not leave a trace of their descendents. How then, could there have been speciation? Many paleontologists claim the trilobite of the Cambrian period is a distant relative of today's crab and lobster. Some trilobite fossils may be thousands of feet deep in the earth and may only be found because of oil drilling cuttings and core samples brought to surface. If these trilobites became extinct during the Cambrian age and left no descendents, and are buried in many thousands of feet of rock, how can we bring them to surface as crabs and lobsters? It would seem logical they were destroyed utterly and completely and suddenly in that age. Annular rings of that age descending to the earth could do this. Again each age is different from the next. Many trilobite fossils are found near the surface.

Who were the descendents of the dinosaur if they had no relatives? They, too, died without apparent reason from a paleontological point of view. Great issue is made of geographic speciation. Yet the facts are, many kinds of animals lived together in the same climate and time period and lived off the same food or each other, in the same ecosystem. This is how it is today. A cow and a horse may live side by side and eat the same food. Geographic speciation had no effect. Cultural, climatic, or geography did not cause adaptation to replace genetics. Where do we find a testable model? There is, however, a wide range possible in genetic variation without crossing or violating the genetic boundary, as proven by Barbara McClintock. Dinosaurs and other large creatures lived from the central United States, to above the Arctic Circle together in the same geological age, as this is one thing the fossil record does prove. They died suddenly and cataclysmically in a tropical and semi-tropical world. Mammoths still have the food in their stomach as proof.

To follow a line of thought in evolutionary theory, are there many kinds of evolution developing different kinds of life independent of the former? It is claimed all life forms evolved from lower forms of life, perhaps from Precambrian times. May I say, there is much that is not yet known about the Cambrian age. If there is no link in life kinds from age to age, are we to conclude that entirely new forms of life developed each time there was an extinction? This will not fit Darwinian evolutionary theory. Doesn't this complicate the theory? Who or what is creating these new life forms not connected to the former? They say extinctions are greater than new life forms can evolve into existence. Dinosaurs, as the fossil record shows, lived together at the same time, in the same environment, and now occupy the same graves, over wide reaches of the earth. I am puzzled, how did they get there? If by evolution and speciation, nothing so far fits together.

If we isolate each of the fossil kinds and ages, they will not connect or blend. In this case, how could we later stir them all together in a common grave? If the dinosaur and crocodilian life were isolated and segregated by nature in the geologic ages and destiny, when was this done? Before death or after? Why then did they all die together in a one world wreck? An interesting article at hand entitled: "Mass Extinctions of the Late Mesozoic". The article discusses the simultaneous disappearance at the end of the Mesozoic age of many reptiles, some marine invertabrates and certain kinds of primitive plants. It seems scholars have sought unsuccessfully to explain the event. There is a new novel hypothesis: The disappearances were the result of a catastrophic disturbance of the biosphere by an extraterrestrial agency. This, however, is right in line with my annular theory though they assign an unworkable agency to the extinction. They wrongly go from catastrophism to gradualism. There is no way the extinction was a gradual process. Every fact is against this theory. They say a clay layer only the thickness of a 25 cent piece edgeways separates the Mesozoic from Cenozoic rocks. These are worldwide ages, not local. If there were zero separation, what would this prove in our hypothesis? How could they prove this? To imply as they do, there was a shut down of the photosynthetic process by volcanic dust in the atmosphere brought about the extinction of dinosaurs, there is a problem. To shut down the photosynthetic process would destroy all forms of life, plant, animal as well as sea life.

Next a theory is offered whereby a supernova exploded above the earth by an encounter of a meteorite with an asteroid, or perhaps either or both with a comet. This cloud of dust would shut down or diminish photosynthesis causing the extinction of dinosaurs and other life. Again, we have the same problem, shutting down the photosynthetic system would not be selective, but destroy all life on earth. This action would throw us way out of the time frame, as scientists say it takes billions of years for life to develop by accident from a single cell. If this process takes us back to zero, that is where we must start with the new life development. It was their choice.

It has been established that sediments of Fort Peck Reservoir of Montana, and areas of North Dakota, coincide with those of Spain. This suggests a worldwide action, not a local action, Mesozoic, or Cenozoic. To establish a crucial boundary by fossil record is not always possible. Paleontologists are stretching this fine line very thin, to prove gradualism and selective extinction in such a gray area. The subject has become a no win numbers game. Cataclysmic annular decline is a reasonable answer to this enigma. The interpretation of geological data which sometimes is not calcuable keeps the issue of gradualism versus catastrophism debated. Both exist, may exist, and operate in succession to each other. The earth gradually built a ring system over millions of years around itself. The rings later both gradually and at times cataclysmically fell to the earth causing great upheavals and extinction of life. Why make something simple so complicated? We may drive a car for fifty years without mishap and one day run into a train. This is both gradualism and cataclysm.

The supernova scheme just will not fit into this scene. Where did the asteroids and meteorites come from? What was the mechanism of arrival for such a flux of material to create a supernova explosions to blot out the sun for a long period of time and not collide with the earth? It would take two objects as a minimum for this supernova idea. This reduces the odds to nil of them colliding in the right place and time. If we use the double-barrel shotgun approach, how could the earth escape permanent damage as to life regeneration? With photosynthesis shut off for a long period of time destroying plant and animal life totally, the oxygen supply would also be reduced. Where would a new supply come from without plant life? Where would a new source of biotic diversity come from? An ecosystem needs a great bacteria world with earthworms. This would be gone. We need some answers.

EXTINCTION OF DINOSAURS

The sudden departure of the dinosaur has had the imagination of paleontologists active for many years. Just a few of the reasons for their extinction are given; they died of constipation. Other creatures ate their food and they starved to death. Plants evolved with toxins to protect themselves and they were poisoned. Climate change is another. Carbon dioxide built up because of a supernova, destroying them. A temperature change came about which caused infertility. An asteroid hit the earth causing a dust cloud to shut down the photosynthesis for about a decade destroying them. The evidence is, however, they died suddenly and not through a slow torturous process. Others claim the mammals did them in. Many more reasons are put forth for their extinction, these will give you a lot of humorous thought. None of these are plausible. Still others claim the appearance of man brought about their extinction. Now then, we have a problem, as dinosaur fossils are not found with man, from the age they live in. I do not imply human fossils are never found near dinosaur fossils. Just seldom. Now, if man had such a high population as to destroy millions of large animals, you would think there would be a few human fossils buried with them somewhere, wouldn't you? Why are dinosaur and other fossils always found on or near the surface of the earth? The claim is many died over 700 million years ago, why didn't erosion bury some of them deep in the earth? If each of these animals lived in widely separated geological ages, why are they found on surface in a more or less common grave? If they assign the trilobite to the Cambrian age and find his fossil there, why cannot the dinosaurs be found deeper if they are descendants of the same? The reconstruction of these animals is according to the opinion of whoever does the reconstructing, they decide what the living creature looked like. Missing parts are invented. In some cases of human fossils, only a piece of bone is found and they invent the rest of the creature to their own liking.

Still another claim is made that dinosaurs some 70 to 75 million years ago, during the Cretaceous Era, migrated from Montana to above the Arctic Circle, raised a family and returned each year. They say there is good reason to doubt the dinosaurs could have lived near the Arctic Circle the year around. Why not is my question, every evidence shows at the time dinosaurs lived, the earth was tropical from pole to pole. North Slope oil drillers can testify to this. The evidence is still under the ice. They also say dinosaurs mated and raised a family then made their southward trek to winter range. They say by analyzing preserved footprints, they are able to estimate how fast the great beast traveled. Do you believe a 70 million year old footprint can reveal how fast the creature was traveling? I am still puzzled as to why the fossils are found on or near the surface, Cretaceous or not? Also, in such wide areas of the North American continent. They must have lived and died right where they are found, is the only plausible answer. They met sudden death as every feature of geology will show; annular down-fall caught them right where they died as also the mammoth.

There must be an answer to some of these questions, perhaps it will be one answer, not dozens. Geologists say the Petrified Forest of Arizona was once buried under 3,000 feet of earth and rock, which was removed by some unexplainable means. We have been taught for many years that when trees or animals were buried they turn to coal and oil. Why didn't these turn to coal and oil? One oil company uses the dinosaur as its emblem. They say these trees were buried for 200 million years. This would place them right in the dinosaur age. In northeast Utah, there is an ancient boneyard of extinct reptilian giant animals which lived they say, 140 million years ago, and were once buried under a mile of earth, and are now near the surface for our viewing. How could these two processes be operating at the same time, one constant and one cataclysmic? Why didn't the reptilian fossils turn to oil as we have been taught they should have? There seems to be great contradiction and inconsistency here.

It cannot be said the biomass pyramid ended its cycle and caused their extinction. No, this would have been a gradual selective kind of extinction, this is not the case. Tremendous tropical and semi-tropical plant life was in evidence at the time of their death from central North America to beyond the Arctic Circle. There is no way we can avoid this triplicity of truth, that, the extinction of the dinosaurs, the end of tropical plants, and a complete change of world climate suddenly, has left its eternal imprint so clear, we can read it as clear as the palm of our hand. If, as we know, the earth had a single climate at that time, how can we accept geographic speciation, due to climate change or differential, when such was nonexistent? Some may jump on the theory of entrapment as an explanation. Now then, with a world so uniformly populated with both plants and animal, what difference would entrapment make anyway? Is there evidence of cataclysm great enough to cause major entrapment of animals, for that matter, even a reroute of a river, which would be nil in affect anyway. The only evidence of cataclysm was at the end of their period on earth, not before. How could they propagate during cataclysm?

With such a universal climate and blend of life, we cannot say either speciation or bioconcentration reduced them to extinction. Yes, bioconcentration is a great factor in our world as man has introduced large amounts of chemicals and poisons into the environment and ecosystem. This was not the case in the pre-deluvian single ecosystem of the earth. Specie extinction and speciation are just not the same thing. Imagine if you can, a world completely free of man-made pollutants and destruction of land. A world with such balance would have no need or reason for speciation, geographic or otherwise. How could natural selection be applicable in a world without need of it free from change. Change is claimed to be the causative factor for speciation. We now live in a world subject to much change, most of it man-made, yet we do not see natural selection working to upgrade life forms. If it is working, it is working backward toward extinction, not upward. Darwin wrote in 1859 in the Origin of Species, dead animals found in one place may relate to living ones, proved evolution. He felt his work at the Galapagos proved speciation, perhaps by entrapment of birds or animals and plants. More evidence against this has been found than for it. An authority on the subject said: *"The exact mechanism of speciation are not fully understood, because speciation tends to be a very slow process."* Some say, change is not possible to detect because the rate of change is so slow. It is not possible to detect the amount of differential over decades of observation. The problem is, if we allow enough time needed to prove speciation, we may be going back farther than the lifespan of the plant or animal in question.

May I suggest an answer. It is not possible to detect a change because there is no major change. There is a wide range in the genetic variation within the specie which never violates the genetic boundary of the specie. I am not trying to put the biologist out of work. Sooner or later, they will have to come to this plain fact. Much more could be said on this subject, though I hope I have given my reader some interesting new thought on an old subject. Because of a tendency toward old traditionally accepted concepts, the truth must always struggle to the light.

BIOMASS EXTINCTION

It is claimed the dinosaurs lived 700 million to a few thousand years ago. The Utah as well as Wyoming find of dinosaur fossils show many kinds of animals lived together and became extinct together. This at the close of the Mesozoic Era. Another theory offered is some dinosaurs changed to birds. Again, there has been no consensus among scientists on their extinction. In all of this we see several claimed geologic processes acting simultaneous and not acceptable to each other. We cannot use the old climate change theory for each as they all lived together at the same time. We cannot fit geographic speciation in here. They say the same kind of climate persisted for millions of years. If the climate remained constant, how could a biomass reason cause their extinction?

In reading all of these accounts of geology I do notice a trend of thought developing. Remarks like this: Upheaval and erosion, mountains were upheaved and eroded, mountain building, rearrangement of the earth's crust, accompanied by uplift, sedimentation stopped as continent-wide uplift increased. This I agree with. It is incomprehensible the power and energy required to uplift and or overthrust continents of the world as has been done. The only competent source would be ocean augmentation which literally squeezed the continents to higher levels. It can only be a delusion to say the petrified forests and dinosaur graveyards were once buried under as much as a mile of earth and rock. I must ask again, why are they found on or near the surface? It is evident some of the bones were washed downstream at this time. The sandstone, limestone, and shales, all show evidence of downstream currents affecting them, as geologists will agree. Everything falls right into place with the proper approach.

This is a quote from one authority on dinosaurs: *"We can only say that the many species of dinosaurs that once existed on earth either became extinct or changed form because they could not adapt to changes that were taking place in their environment. The mystery of dinosaurian extinction may never be solved.''* This admission leaves a missing link in the evolutionary chain. Paleontologists keep digging in their quest for knowledge of the past. Looking for windows which will allow them to look backward for answers. The same answers keep coming up. We need a new premise to work from. Continental upheaval by ocean augmentation from annular systems is a place to start anew. Some dinosaurs such as the Brachiosaurs weighed over 50 tons. The Nanosaurus was one of the smallest, about the size of a chicken. Between these two extremes many kinds of dinosaurs and reptilian animals existed side by side in the same time frame. In this spectrum there is no room for geographic speciation. Animals of this time did not die off gradually as evidence shows, but they all died suddenly, cataclysmically. The ecosystem they lived in was a real going concern when they became extinct. Both herbivores and carnivores lived together and are found together in the graveyard of nature. There is no evidence of slow extinction. Teaming life was a going thing right up to death and this on a worldwide scale, not just local. Many kinds of fossils including mammals were found at Como Bluff, Wyoming fossil find. To say carnivores first destroyed to extinction the herbivores, and then starved to death, cannot be the case. Most scientists and paleontologists agree the herbivores always outnumber the carnivores. Most agree, microorganisms and plant food provide energy first to the herbivores which pass it to the carnivores. In this biomass, indications are, there is an interaction with dinosaurs and all creatures of the time.

A biomass-ecosystem is a very interesting entity. Say we had a square mile of land called a section (640 acres). Let's say the plant growth on this section will support 100 head of cattle in perpetuity. Then the land owner becomes greedy and puts 1,000 head of cattle on the section of land. Can this land support 1,000 head of cattle? Yes, for a short time, though over a long period of time, the biomass-ecosystem will be destroyed and the cattle will die off or down to what the land can support. Because the biomass cannot pyramid forever, or invert, we must sacrifice perpetuity. Perhaps if a few cattle survived, they could again produce up to an acceptable number — this is another subject. The energy source which is solar has not changed, it is perpetual. There is only a short term gain when a normal ecosystem is tampered with in its ability to produce. This is true when a very large dam is built. Energy levels will not move uphill unless an outside force is applied, which may be limited. We must apply this to the entire earth. Dr. Paul Erhlich said; *"We could become extinct.''* We would agree with Dr. Carl Sagan when he said: *"One generation should not own the earth.''*

Energy and environmental laws of the ecosystem apply to large dams. I do not mean to say no dams should be built. Many are badly planned and designed, or in the wrong places. Many should not have been built. Many dams destroy more behind them than they produce in front of them. Perpetuity is traded for a short term gain. Like going on a spending spree, retribution will never leave us. All dams will silt shut in time as a negative factor. We traded perpetuity of production for short term gain, and greater loss financially and ecologically. Men, as intelligent creatures should learn, at least from experience.

This reminds us of the 640 acre biomass. Have you ever thought of flood control, that there is no such thing in reality. Flood control means destroying many others to save a few. It may take 10-40 acres to save one, depending on where. The trade is out of balance and we must give up perpetuity in the process. A town may build dikes to keep rain water from running through the streets when it rains. A hard rain may come one day, the dikes break loose and floods the town with much damage. This would apply to farmland as well. If farmers build dikes on both sides of a river it will raise the water level higher in between and flood the people and towns in between. I am not saying no dikes should ever be built. A few well designed may help, most will not.

Most dams will destroy the ability of the biomass to produce, not increase it. To flood land to save land from flooding leaves the ecosystem short of energy and production. If a dam could be built where the land it will flood is of very low productivity and the land it will save is of very high productivity, there is a gain. The reverse is usually true, the very best land is flooded. In late 1981, CBS-TV News had a series of newscasts showing many dams are nothing more than "pork barrel." Francie M. Berg, in her interesting book, South Dakota — Land of Shining Gold, made some interesting observations. *"Many ironies can be found here. Indian people see a bitter humor in this typically 'white men solution': Flood land to prevent flooding. Land underwater never floods. Permanent flooding of more than 80% of South Dakota's Missouri bottomlands seems a brutal way of preventing the floods."* Another quote from the same book. *"A local joke asks the question: 'How long will it take for the dead trees to disappear?' The answer: 'About as long as it takes for the dams to silt in.''* Francie M. Berg has a number of interesting books. This would apply to Garrison Dam in North Dakota. Many people have suggested a series of smaller dams would be far better.

One of the best examples of failure is the mighty Aswan Dam on the Nile Delta of Egypt. Egypt's Institute of Astronomy and Geophysics in Cairo reports much of Egyptian heritage and cultural structures are in danger of collapse. The report says a sea of underground water from the high Aswan Dam has raised the water table within 6.5 feet below some structures. The pyramids are in danger because of floating on a sea of underground water. Other structures in danger of collapse are the Spinx at Giza and the Temple of Karnak. The greatest destruction, as news reports say, is the ecology with the Answan Dam.

Perhaps the oldest irrigation project on earth was in Babylon in the Tigris Euphrates Valley in the B.C. period. Salinity laid it to ruin. The Soviet Central Asian irrigation project is an evironmental disaster in the making. Mudflats and salt pools are endless according to the Los Angeles Times, Sunday, March 1, 1981.

There is the problem with the over pumping of the great aquafers of the earth, to raise crops in overproduction and government subsidy. These aquafers would supply domestic water for the rest of eternity if not wasted on crops in oversupply. The economy is as bad off now as the ecology from man-made causes. One man's overhead is another man's living. Water evaporates while salts and minerals do not. We must learn to live with the biomass-ecosystem or perish. Natural farming in a fifty year period will far outproduce centralized intensified farming where a large part of the land is used for the project. The evaporation factor is the greatest enemy of large scale irrigation projects with seepage not far behind. Add overuse of chemical fertilizers and success is reduced to nil, in long term. There is no way we can violate the biomass-ecosystem. If we add an outside force we lose perpetuity of the system. A 747 jet plane can climb to 40,000 feet and travel 600 mph in a neutral realm until the outside force, fuel, through the engines runs out. It must return to earth for a new supply, there is no perpetuity without an outside force. Thermodynamics say a reaction will not go uphill without an outside force. Slogans say: use it or loose it, even if you don't need it. People have gone broke buying things they do not need because they are for sale at a bargain price. When something is objectional and uneconomic, propaganda must be used.

Dinosaurs did not become extinct because the biomass-ecosystem failed. Man is causing the extinction of life and could become extinct himself as a result.

CHAPTER 17

EVOLUTIONISM — CREATIONISM

"In the absence of any other proof,

As I have said before, this is not a religious book or Bible book. (Religion and the Bible are not the same thing.) For historical and other reasons, I will base the following comments. Because the evolution — creation subject has been injected into our historical times by men of science and religion, I would be remiss in not discussing the subject in my theme. Many of my remarks will be based on the work of Dr. Carl Sagan, and Dr. Stephen Jay Gould, Cornell and Harvard Universities. Many others could be quoted; however, these two men are prominent at this time. Also Dr. Paul Ehrlich of Stanford University. I have gained much valuable information from these scientists and many others. I will remain neutral and not take sides with either the evolutionists or the creationists. From this position I will be better able to look at both issues in a more objective way. Another reason I will remain neutral, both sides are political. How can a court of law decide right or wrong in an issue where both are equally in error?

First, it has never been established which is religion and which is not, in the two issues. Discover Magazine for January 1982 discusses paleontologist Stephen Gould and his religious fervor in opposing creationism. Where is the line of demarcation, where the one side is purely secular and the other is not? Are they not both hierarchial? Isn't it true, the entrenchment of ideas usually becomes hierarchial, with a few at the top setting the pace and taking the lead in worshipful manner? This was true from Copernicus to Galileo. The coin has just been flipped over and science was on the other side. We are all thankful for the factual scientific knowledge from Kepler to now. It took fifteen hundred years to correct Ptolemaic theory. How far removed from this are we today? Isn't science just taking the hierarchial position religion once held, and still does in some fields of human thought control (right or wrong not being the issue). Isn't medicine today a hierarchy? A hierarchy requires a priesthood. Don't we see a hierarchy with its priesthood in each of these areas today? Isn't it true, that the witchdoctor is a hierarchial system? We are all expected to follow without question.

Frankly, if we took all evolutionary science and all religious doctrine (not biblical) developed in the last one hundred years and discarded it, wouldn't we still be in the same intellectual void we now are on these subjects? Evolutionary science makes wide sweeping claims not provable. I am not a member of a hierarchy, yet I absorb the same information and come up with an entirely different set of ideas. Why is this? It just depends on where we started from, not where we ended up. A mile is always a mile, in a race we must all start at the same place. Science will argue, there is no possibility of even a single virgin birth. On the other hand, they claim billions of billions of trillions of creatures and life forms over billions of years developed from one single accidental cell in a primeval sea, in virgin birth style, without due process of biological reproduction. E. O. Wilson of Harvard holds that much of human behavior is directly controlled by genes. Gould, also of Harvard, disagrees, calling it "biological determinism." Why shouldn't they agree? Can you imagine the captain and copilot of a 747 jet plane, in an emergency, disagreeing with contents of the flight manual? We might argue, an arrow needs a designer, DNA does not — it's an accident.

Discover, same issue says, *"Dr. Gould is an expert on Ontogeny (life cycle of a single organism) and Phylogeny (evolution of a group)."* Part of the claim of evolutionary science and paleontology is as the article says they must attack the big problem — how does evolution occur? I thought they said it already occurred!! Evolutionary biology argues that there has been a grand sweep of evolutionary change. They feel the fossil record will eventually prove this. Why would fossil genetic evolution go on for millions of years and suddenly jump great time spans is next to impossible to prove, let alone understand. The theorists would sure save a lot of time if they can sell the idea. They no longer would need to look for the missing links. Just find a bone here and one there and fill in the information as they believed it to be. There is no fossil connection between the geological ages. This would be an easy way to cross from one to the other without accountability to the fossil record. The Precambrian is a worldwide geological age. It is about 16,000 feet deep in the Williston Basin in places. I have mentioned this before, it is called the deadwood sand, as it outcrops at Deadwood, South Dakota. It would be easy to solve the fossil world in this basin if we adopted the new theory of sweeping jumps in fossilization. Perhaps the differing groups could go to a court of law to settle the issue. Would a judge in a court know the difference between a Cambrian trilobite and a dinosaur? Paleontologists say they are cousins or whatever, though each left no relatives behind.

The clash between the evolutionists and creationists will produce some good, in that they will each expose the other. Perhaps some facts will be left over to consider. The disaster of the world is centralization — economic, government, religion, social and science. Can courts settle everything? It is possible to legislate against crime, though it is not possible to legislate righteousness. If it were possible to legislate righteousness, there would be no need for the law against crime.

It would seem evolution is another religion. They are fragmented and divided among themselves just as religion or creationists are, though each maintains a hierarchy. How can a court of law decide this issue, it must have a basis for judgment? Creationists claim the earth was created in seven literal twenty-four hour days. Six if you prefer, as God rested on the seventh. They give the evolutionists a lot of good ammunition in this claim as it is not scientifically provable, nor is it biblical either. The creationists may argue; God is all powerful, so He could create heavens and the earth in six or seven twenty-four hours days. I will not argue with that point of view, but that just is not how it really was done. The evidence is overwhelming that the earth is billions of years old. The Bible does not say the earth was created in six days. The Hebrew words used in Genesis allowed for days of many thousands of years. A court of law only has one stand to take as it cannot rule on religious doctrine or evolutionary belief, both being wrong and unprovable. The court will rule on constitutionality of separation of church and state. This is not to say that church and state are really separated. It appears religion and politics have been mixed together for centuries.

We must look at many issues in total concept, in the broad theoretical picture. To build on this subject some historical information will interest you. Religion (not Christianity) has been around for a long time. So has the issue of separation or union of church and state. The greatest union of church and state the world has ever seen was held at the Council of Nicaea, in A.D. 325, by Roman Emperor Pontifex Maximus, Constantine. He fused most of the religions of the world into one. Nicene and Athanasian Creed came from this Council. As one historian said, *"This set the stage for worldwide bloodshed til our day."* The 30 Years War of 1618-1648 in Europe is one example. Roman Emperor Julius Caesar increased the number of pontiffs in the college to sixteen. Under pagan Emperor Tiberius, all sixteen of the pontiffs were pagan. Augustus Caesar held the title Pontifex Maximus for the last part of his life. Union of church and state were strong. The world empire of religion is based on the Platonic idea that all souls are immortal. Add to this the hellfire doctrine and a few other pagan teachings, plus hierarchial power and it becomes an empire. Courts within it would be its tool.

William Tyndale learned this when he was condemned as a heretic in August, 1536. The same year he was strangled to death on a stake, and his body burned. What was his crime? He translated the Bible into the vernacular of the common man. He wanted even the plowboy to be able to read the scriptures. Clergymen and theologians attended his trial and approved his condemnation. There are thousands of kinds of religions, some even claim the Bible as their manual. We are all very glad the flight crew on our next plane trip do not treat the flight manual this way. I am making a point as you know. The word heretic is now used among the evolutionists in referring to others who do not accept the ideas of the evolutionary hierarchy.

The same kinds of organizations are still with us today. I am sure nobody will be burned at the stake as the battlelines are drawn between the war of ideas in creationism and evolution. They are both fundamentally wrong so what will be the basis for decisions? Who will make them, what will the judge use for testimony of proof? Perhaps the deciding factor will be the triumph of propaganda over truth, with theatrical speculation adding to the winning side. There have been many great men in science over the centuries, usually coming from the grass roots. Will true science be destroyed by jealousy, competition, bigotry and fear of financial and leadership roles? Isn't commercialism their main objective now? Some feel science has already created a Frankenstein ready to devour us all. It is already beyond the point of no return. There seems to be no other objective to gene splicing but commercial gain, not life enhancement.

When has truth and fact ever been popular? It usually is the last to be accepted, and this only after great pressure overwhelms the falsehoods established by the control of the propaganda in favor of the false ideas. Why are there thousands of thousands of kinds of religions? Shouldn't truth reduce them all to one? In discussing the subject with a man one time, he said, *"No matter how many religions there are, there always will be one more, as I have my own religion."* What is more, he said, *"I am not seeking any converts to my religion as it is my own, I am the clergy, laity, and total membership. I would not welcome any converts."* The point is, both religion and evolution are about the same. They are fragmented into many opposing factions within themselves and are not able to arrive at truth. Thoughts and ideas which do not conform to the status quo are suppressed, or not given to the public. This was true with Copernicus, Galileo and Einstein. It is not any different today. Vegetarian geology is not true, yet it has been accepted for centuries. History will show millions of people were tortured and killed during the Dark Ages because they had a different viewpoint.

Perhaps a large part of the misery on earth, over the centuries, was caused by the idea and philosophy of the Divine Right of Kings. In this way, laws on treason could be expanded to cover both political and religious, for a broader control of liberty. It would seem Sagan and Gould are correct in an assessment of religion, being only mental regimentation. Competition and jealousy are prevalent in high authority. The month of July is named after Julius Caesar, it has thirty-one days. August gets it name from Augustus Caesar, which only had thirty days. This small fact seemed to bother the great Augustus, so he took a day from February, adding it to August making it thirty-one days also. In a very educational and interesting book by Dr. Carl Sagan, "Broca's Brain", you may agree with much of his reflection on religion. If you have read, or do read, this book, you will be better informed by so doing. Also, a book by Dr. Stephen Jay Gould, "The Panda's Thumb". I will not rehash these books, though I will comment on them. These and many other books available will add greatly to your storehouse of information, though they will not answer or solve many of the great questions they address.

Many of these authors use, as a basis of refutation of God, quotes from high religious authorities and religions of long standing acceptance. They show how wrong and inconsistent they are, which is true. This is where I disagree, as these religious men do not represent God or the Scriptures. They speak for their philosophies only and where is their base from which to project from. Mainly from Dante, Plato, and Socrates. These and many other men were only historical philosophers. Dante developed the hellfire and torment doctrine. Plato the immortality of the soul. Socrates (469-399 B.C.) left no writings of his own, and is known only through the writings of his pupil Plato. The many statements confidently made by theologians are based on Greek Mythology — what does this have to do with God? Sagan says in his book, Broca's Brain, *"It is a curious concept this, of an omnipotent God with a long list of things he is forbidden to do by the fiat of the theologians."* As Sagan says, *"Religion is fundamentally mystical, the gods inscrutable . . ."* Quote from Broca's Brain: *"Certainly, bureaucratic religions have throughout human history allied themselves with the secular authorities, and it has frequently been to the benefit of those ruling a nation to inculate the faith."*

Voltaire argued, that if God did not exist, man would be obligated to invent Him. As Sagan asks, *"If religions are fundamentally silly, why is it that so many people believe in them?"* May I add, the reason is for mind control — they are not from God. Some feel if scientific and mathematical knowledge had developed earlier, religion would not have had such progress. For example, Galileo discovered four moons on Jupiter. His theological contemporaries held to the idea that other planets did not have moons. This, no doubt, was upsetting to fundamentalist churchmen of the time. Some persons may argue, this shows the Bible was wrong as the theologians claimed to read it and were wrong. This shows the opposite, as the Bible did not say only the earth would have moons. The fact that the earth had a moon would seem to indicate other planets would have a moon or moons.

Many are those who would make an issue of the seemingly impossible events from the scripture, one being the earth's rotation stopping in the Book of Joshua. Dr. Sagan covers this under the earth's rotation. I agree with Sagan's thoughts in physical law. This would create two problems, not one. To stop the earth from rotating would be a disaster to everything and everybody on earth. To start it rotating would be equally as disasterous. From 1,000 miles per hour and back in 24 hours is 2,000 miles per hour factor wise. As almost any intelligent person would know, the Bible uses much in a symbolic way. It just does not say in the Joshua account that the earth stopped rotating. There is another logical answer for such seemingly phenomena. We could very well say that stars, moons, and lights from the heavens, may well picture the rulers and men in high positions as well as scientific and religious, have led the world into such darkness, in a symbolic way, these lights might as well not be shining. We now live in such a darkened world. These great stars do not give off light any longer. Could a court of law decide these issues? Most of them harbor on the brink of lunacy from a religious and scientific point of view.

The Greek Court of the Areopagus had the power to fine or imprison citizens for almost any offense, no matter how trivial. They could sentence people to death if the court disagreed with a new idea on any subject if they so desired. The Apostle Paul was dragged before the court of the Areopagus to give his views. His opening words were, *"Men of Athens, I behold that in all things you seem to be more given to the fear of the dieties than others are. For instance, while passing along and carefully observing your objects of veneration I also found an altar on which had been inscribed 'To an Unknown God'."* At this time, the court was under Roman rule and the power of execution had been taken away from it.

Another court of interest was the ancient Jewish Sanhedrin located at Jerusalem. The history of the court is not clear; however, here are some of its features. It seemed to be well structured in favor of the victim being tried, though any court may be corrupted as this one was at times. It ruled on civil and religious matters. The court had seventy-one members counting the high priest who presided. This office was appointive under Roman rule; political then. The high priest and chief priest are not the same. High priest Caiaphas, presided at the trial of Jesus, Ananias presided at the trial of Paul. A quorum of 23 judges was required to act on a case. If twelve voted for acquittal and eleven against the victim was freed. If a judge voted for acquittal, he could not change the vote to condemnation. On the other hand, if a judge voted for condemnation, he could change his vote for acquittal. The court had two clerks, one for acquittal and one against. A two-thirds majority was required for condemnation, and a simple majority for acquittal. The Jewish system also had lower courts, some of three men and some of seven. Man has experimented with many forms of jurisprudence. The territorial range of authority must first be determined for jurisdictional control. Justice is usually difficult to come by. In A.D. times, under Roman rule, the court could not rule on capital cases without confirmation from the Roman Procurator, though usually ruled in accord with the Sanhedrin. The Roman Administrator granted the power of death without confirmation in cases where a gentile crossed the dividing line of the Inner Court and Temple. Voting was done by judges dropping a black pebble or ball into the box for condemnation and a white one for acquittal. Its power did vary at different times according to politics. The high priest presided over the seventy-one member religious court. Jesus was condemned by this court, though all concerned seemed very glad to let the Romans carry out the execution, just in case Jesus did turn out to be the Christ. The Sanhedrin disappeared after the second fall of Jerusalem in A.D. 70.

With all its good structure, the Sanhedrin Court became a tool of those in control of religion and government. Church and state was not separate. We may only wonder what such a court would do with the trials of evolution and creationism. We may think back in history and wonder how men like Adolph Hitler came to power. What had happened to the courts and parliamentary systems of the world? Men not close to such may wrongly just dismiss him as an upstart party leader. He was aptly described by some as "a man with a cork leg throwing an epileptic fit on a tin roof." As Pierre Van Paassen said in his book, "Days of Our Years", the German people are too soberly analytical to take him seriously. Yet, the stranglehold hammerlock was applied to civilization by Hitler. Some have said, freedom is never lost, it is thrown away. History repeats itself because it is lost to the next generation. Where was religion and atheism when all of this was developing?

SOME REFLECTIONS

Sagan says, *"Much of history is a painful liberation from provincialism. This deprovincialization has been greatly aided by space exploration."* Now that it has happened who is any better off, except those who live off the process? We do live in a parasitical system, don't we? In the planetary differences, the Darwin insight is that life elsewhere is fundamentally different from here. Even if this is perceived by many in what way will this unify the human family? To the Greeks of old, anyone not a Greek was a barbarian. If technology advances cause us to realize we share a common life raft, it has not shown as yet. The earth is still dying an ecological death with each person expecting the other to save it while they draw on its resources.

We may read the Mismeasure of Man, by Stephen Jay Gould, this book is full of interesting information, and arrive at one simple conclusion, "regional adaptation within a permanent genetic boundary."

Next, read The Panda's Thumb, by Gould, where you read many remarks like this: *"The geological record is extremely imperfect." "Darwin's argument still persists as the favored escape of most paleontologists from the embarrassment of a record that seems to show so little evolution directly."* Next, *"Only an immense span of time had permitted such a sluggish process to achieve so much."* For forty years I have read books on these and other subjects. My study is full of books. I built more shelf room and filled them. I bought more bookcases and filled them. I have books piled on my living room floor. I am still buying books and read with fervor when some great paleontologists say there has been a breakthrough and evolution can now be proven. As I turn each page, I feel I will have the answer on the next. No such thing has ever happened. I feel like I have been lead on by deception.

The same is true of finding life in outer space. Thousands of words, but never any proof of life out there in the great universe. I saw all of the Cosmos shows once and some a second time and read the works of Sagan and others, all with extreme interest, and still have not found any evidence of life, extraterrestrial life. Not little green creatures, not a fungus, or even a microbe. It is a total void, nothing. I see no reason to believe man will ever be able to do much in space without a robot or machine to go out there. We are locked, inescapably, to the organic world and, above all, to oxygen. Unless this kind of environment is developed on another planet, we are pretty much stuck here on earth.

It appears science today is just as iconoclastic as religion has been, in that they are all lined up in battle with double-barrel shotguns, shooting down each other's ideas hoping for a personal breakthrough and becoming another Einstein. Please do not feel I oppose evolution for religious reasons. I do not oppose evolution as a science. I only disagree with their claims. From their own position, they cannot prove their own claims.

There has never been one proven case of evolution by speciation. Life has not been upgraded by accidental chance mutations. After tens of thousands of pages of material from the best scientists and paleontologists, as well as astronomers, I do not regret the time spent. These have added immensely to my knowledge for which I could not arrive at some final decisions. Gould says (Discover), *"We must combat the*

*few yahoos who exploit the fruits of poor education for ready cash
and larger political ends.''* He says, *"Evangelical rights are used to
establish a much more authoritarian and religious-based society.''* I
am not now, or ever have been, a fundamentalist. The dark ages was a
society of authoritarian religious domination. They have lost their
power though are still around. Could he (Gould) mean Billy Graham
and other TV preachers, like the Moral Majority? Could he mean the
Pope of Rome or does he exempt him from this. Perhaps the Protes-
tants!

If science will suffer from a religious society, they are determined
to see that it does not happen. Science, too, must worry about its
ready cash. Much of it is taxpayer or direct government subsidy
money. Each faction of society feels justified to keep its hand in the
cookie jar. Do the evolutionists feel all religions of the world should
be dismantled? A prominent communist leader once said, *"Religion is
the opium of the people.''* I will remain neutral, however.

Perhaps we should accept everyone at their face value and by
what they say. Evolutionists, atheists, and scientists imply, if a completely secular society could be developed this would solve the world's
problems and we could live in a paradise. It is now an accepted premise by many that the most intellectual educated people on earth have
been in charge of government, religion, science, cultural, and all en-
deavors of man. There is the United Nations to regulate world affairs
peaceably. Yet, all through history the secular society has been tried
and failed just as the religious controlled governments have. Perhaps
Voltaire was right when he said, *"If God did not exist men would invent Him.''* The social regime of Stalin was a secular experiment
which failed. Pictures and statues of himself and other leaders were
deified and placed everywhere, just as with Mao Tse-tung of China.
They invented a god. In China Neo-Confucianism was developed as a
replacement for religion. It turned out to be nothing more than a prag-
matic metaphysical speculation. It seems to impede and frustrate
rather than build. Modern communism is so proliferated no one can
recognize it any longer. As to the success of a secular system, for anyone who has the stomach for it, read "The Gulag Archipelago, by
Aleksander I. Solzhenitsyn." (Harper & Row)

In my numerous visits to the United Nations Organization, I got
the impression it is a purely secular organization; what else could it
be? I once sat in the General Assembly Hall and reflected back on all
of the world leaders who spoke at the rostrum, from Nikita Khrusch-
chev who pounded his shoe on the rostrum and the Pope of Rome.
What are its accomplishments we may ponder? On the Plaza across
the street from the U.N., apparently in philosophical way only, the

words from Micah 4:3 or Isaiah 2:4 are carved in stone, in part: *"And they will have to beat their swords into plowshares and their spears into pruning shears. They will not lift up sword, nation against nation, neither will they learn war anymore."* From the time these Biblical words were placed on this plaque, until now, the reverse has been true with the educated intellectual leaders of the world.

The only alternative I can offer is the Millenial rule of Christ.

As a young man, the first large museum I visited was the Field Museum and Science Building in Chicago. I was awestruck with dinosaurs and all the things I had never seen before. I visited it several times and must go back again. Also, the American Museum of Natural History in Manhattan where I have visited several times. Also, the Smithsonian Museum in Washington, D.C. When we travel, I stop at every museum I come to, no matter how small. I even love old car or tractor museums. My first plane ride was at Minneapolis where a friend and I paid a few dollars for a ride over the city. I was awestruck with the machine. I made up my mind to learn to fly as soon as I had the money, which I did. An oil drilling rig always fascinated me as a young man and decided some day I would drill one. As it turned out, it was my privilege to supervise the drilling of many oil wells. How would I describe myself? Perhaps as Discover Magazine described Gould's father, *"A self-taught man, an intellectual without official credentials."* Just for curiosity, I counted the books on just one shelf of my study and made an estimate of the average pages in each (Flavius Josephus over 1,000) and my calculator shows 50,000 pages and I use them as reference now. My library contains many more books.

It may appear I am blowing my own horn; however, I am not a graduate of Harvard University or any other university. When in the business world, I employed people from the universities. Not to cast a reflection on Harvard, though a lawyer we once employed who was a Harvard graduate would get you in more trouble than he could get you out of. Another time, in some business association with a Harvard graduate, it was a sad experience as he had no business judgment at all. Many people I know with little or no education had very good business judgment and were successful. Why is it most of the millionaires I know do not have a college education?

Perhaps it will be necessary for me to blow my own horn, as I do not have Harvard, Yale, Cornell, or Stanford University behind me to promote my ideas. The problem with higher education (though necessary) is there is a tendency to mold the mind into a certain line of thought which is hard for a student to ever break away from. Again, I do not compare myself to Einstein; however, I am an independent thinker as many have been. This does not mean I do not need to consider ideas from others; no, the opposite is true. In most projects, regardless of the nature, I have found myself in conflict with the entire chain of thought development. Yet, I do not have difficulty in arriving at profound and absolute decisions. Without the contributions of others to the overall picture, this would not be possible. The entire social structure we live under today is bent toward making us a product of conformity. True, many mechanical and scientific, as well as economic objectives, may be standardized to our benefit. Others may not, in fact, this is where we may become victimized. Too often progress has been halted by such standard regimentation of ideas and thought. Preconceived opinions may stand in the way of progress. Some changes are detrimental, yet the ecological survival of the earth requires a change from what it now is.

CHAPTER 18
LOOKING BACKWARD AND LOOKING FORWARD

REVOLUTIONS IN SCIENTIFIC THEORY

Because of regimentation of thought progress, there have been many revolutions in scientific theories. Intellectual integrity and consistency is not popular in science if it does not support the status quo of science at any given time in history. This was true as far back as the Greek Scientific Era 600 B.C. Hasn't the scientific realm of today grown powerful and acts much as the Church did in Galileo's time? Anything can become a religion as modern science is today, as well as any movement political or otherwise. Communism can be a religion just as Divine Right of Kings was under the Greek and Roman Empires. Many ideals based on human authority are such. Rulers today claim Divine Right. Einstein once said, *"We are not as far removed from Galileo's time as we would like to think."* Many theorists have been set in their own ways and dogmatic as a defense against criticism or being wrong. New ideas were not accepted unless they conform to a given set of rules decided by a few at the top (a hierarchy). Never should we avoid information and ideas which may cast doubt on our accepted premise on a subject. We should seek such information and even wish we would find a point which would disprove our own as a test to our convictions. Then, if our position cannot be shattered we know we are on solid ground. If then our convictions are reduced, we are learning. If our convictions are enhanced we are learning and moving in the right direction. When a field of science or evolution becomes a fad, a dogma, and is hallowed, it is time to sit back and take a long look at it. Emotional conviction did not place a man on the moon, or send the Voyager Missions into space. Much scientific theory today has become a creed, not a fact. It takes a great deal of faith to believe a lot of it, just as it takes a lot of faith to believe in some mysterious man-made religion. A number of experts have said: It is an extraordinary act of faith to believe in spontaneous development of life by chance. Many use the same data to prove absolutely opposite theories.

Everybody, it seems, quotes Einstein and Abraham Lincoln to prove their ideas. Both of these great men believed in God. History shows Lincoln did not belong to a church and was a Bible reader. I am not sure if Einstein had a formal religion or not. They had humility and modesty and were successful, while many of those who quote them do not, and are not successful. Someone said, *"Humility and modesty are the weakest of all human traits."* Humility, not self-confidence, is a great virtue as expressed by renowned scientist Albert Einstein. He said, *"Everyone who is seriously involved in the pursuit of science becomes convinced that a spirit is manifest in the Laws of the Universe — a spirit vastly superior to that of man, and who in the face of which we with our modest powers must feel humble."*

Revolutions in science often are introduced by itself. Often, they strain at a gnat and swallow a camel. Theories may only introduce an essential of what can be experimentally observed — only the concept of stability. On the one hand, science will say they absolutely know everything. Then, on the other hand, they will say: there are no absolutes. It is not possible to make something correspond to Einstein's special relativity if it does not. There is a propensity to transform all concepts into orthodoxy. Orthodoxy sets up an intellectual rigidity not supposed to be acceptable to science. In a discussion of quantum mechanics (energy is not infinitely divisible, an assumption) Einstein had as a counterexample, he would say, *"God does not play dice."* (German Scientist Max Ernst Ludwid Planck — 1858-1947 — discovered an entirely new field of physics, known as quantum mechanics. The theory has to do with constant and frequency.)

It appears there is an almost childish jealousy and vicious competition in modern science and technology — include evolution. The issue of constancy and permanency is still unresolved. Isn't constancy found in frequency (electric cycle, wavelength) as much of the universe is locked to permanency by constancy of frequency? If deviations are constantly deviatory so as to build a premise on them, who is to say which are norms and which are not? Wavelengths are always constant yet moving. If this were not true it would be impossible to transmit television pictures from Mars to earth.

Until proven otherwise, all life must have been a direct creation. Darwinian evolution and modern evolutionary science, predicated on Darwin, are on their last leg, in a balancing act for survival, based on geographic variation, which is not speciation as they claim. The problem with a balancing act, is, it is all chance and may fall either way. So it is with the evolutionary premise of too much relevance on small variation in geographic variations. If minor variations and accident are the stuff of evolution, and if we add all small effects through a long period we have no gain or loss. If we recognize this principle as a basic mode of reasoning we must not deviate. Like flipping a coin ten thousand times or ten billion times we should about break even on the number of heads and tails that turn up. We will grant minor variations on either side of a line in genetic variation; but, where do we find a gain or loss in favor of speciation in the long term? None is apparent! No matter how far we may carry this continually recurrent cause it will exemplify the fact, that the process of genetic evolution has been rigged and not constructed for modification in a great degree. (Rigged by whom, is the question?)

One hundred years ago Darwin's theory of natural selection was considered a revolution in scientific progress. Proving it wrong in our time would be considered a revolution in science. Darwinian evolution has never been acceptable to my mind, even as a boy. Not for religious reasons, but rather for scientific reasons. It is nothing more than a philosophy, a faith in the unknown. Personal preference and sentimentality will neither add nor detract from the known framework of boundaries we must remain within. Time has been thought to be the salvation of the evolution theory. Geology has never allowed enough time as they say, to allow observable genetic speciation. The creature always becomes extinct before change occurs. It must be maddening that these creatures should fool scientists by having no forefathers or relatives. Darwin's natural selection is not speciation, nor is adaptation. Why must order be only the by-product of struggle? The weak always outnumber the strong so all become weaker, even the scientists.

As one scientist suggested, *"We must develop criteria for inferring the processes we cannot see from results that have been preserved."* To use this method to project backward into time as an accurate scale to measure from we must have constancy of factor. Do we have constancy of factor? No, we do not! We are referring to millions of years even billions. If time is the missing element in proving evolution, it may never be proved as time destroys, not confirms, evolution. We cannot infer the unobservable past by the observable present without indulging in outright speculation, so felt Sir Charles

Lyell (British scientist and father of modern geology, 1797-1875) who felt small change over immense periods of time was the only answer. "Uniformitarianism geology" thus came into being. It has long been my contention that present known processes which lack constancy could not build our geologic world. If we are to add the total sum of continual causes, as some dust here and some erosion there, we run out of time as usual — this will be true of genetic variation.

Ask any geologist after seeing an oil well core sample to surface from the Cambrian, if he thought it had a comparison to the bacteria and worm developed soil on surface today in an ocean of free oxygen? Am sure no geologist would say the Cambrian age was developed by a uniformitarian system which worked its way to the earth's surface as we know it to be now over millions of years. How can we apply uniformitarianism to something as spectacular and radical as the development of the Grand Canyon? The observed results are too spectacular to be built by a little dust blowing around and some erosion by water. If time is the factor which built the Grand Canyon, there would be no canyon at all, as the banks of both sides would have been affected the same way as the canyon area was. Water did not run uphill for millions of years. The Grand Canyon was built by sudden cataclysm by the annular process. All cataclysmic processes must become dormant or seek ultimate dormancy, or the world would self-destruct. The fantastic world has not been constructed by the sum total of almost unobservable variation over billions of years of constancy and uniformitarianism. Why then, is it heterogeneous and conglomerate with millions of life species? The geologic ages and stratigraphic column reveal great variations which are easily observable. This has not been left to chance and guesswork by creative process.

Enigmas Of Evolution, is the title of an article in Newsweek, March 29, 1982. This article discusses some interesting aspects of evolution and names a number of prominent persons, mainly Dr. Stephan Jay Gould of Harvard University. There is not unity among the scientists as is shown in the article.

The revolutionary departure of evolutionary thought will spell its demise in our generation. Evolution will be hung on the very scaffold they have built to hang the creationists on. Its death will have been brought on by itself because they did not have discernment. The saga of evolution will end sagaciously at their own whims. The inability of science to find any transitional forms of fossils to complete the lineage of ancestors after hundreds of years of digging has made their absence a glaring sore thumb. This problem seems unsurmountable and is the foundation, as they say, of their theory. Gradual transition over long time periods has been the main stay of evolution since Darwin. Evolution is a dead end today.

For years I have wondered how they would cross this great obstacle in fossil transition which has defied explanation. They had to change the rules in the middle of the game when they found they were losing. They finally came up with a theory which will not set well with intelligent people. If they could not find the missing links they would get rid of them by by-passing them. Here is the explanation they give to avoid this glaring missing link problem: *"A new theory of evolution, instead of changing gradually as one generation shades into the next, evolution as Gould sees it proceeds in discrete leaps. According to the theory of punctuated equilibrium, there are no transitional forms between species, and thus no missing links."* This theory is laughable to all thinking people. This theory gives 100% support to direct creation of life forms.

This theory of no missing links completely destroys the credibility of old-school evolution. It not only attempts to get rid of the missing links, it gets rid of the entire chain. Sometime try to pull a stuck car out with a chain missing links. The greatest enemy of evolution has always been the idea postulated by paleontologists of great time, such as eons and epochs. Let us use their own position on evolutionary and geologic development. If we had a legal measurement for minute deviations in genetic change or geologic change or any other change predicated on such, after running it through a computer we are driven completely off the time frame which they call about four billion years. Now, to assert that species arise relatively fast is ridiculous from their own position. Scientists have played around with the poor little fruit fly for years and proved nothing in their favor. To now tamper with such an entrenched idea as they have stood on for so long and totally change this mechanism seems irresponsible at best — with less supporting proof than the former position. The evidence is strong that it is time to abandon the entire mechanism not repair it. This new theory is tantamount to saying, there is some good in the use of nerve gas in that it stops tooth decay.

If extinction is the common fate of all species as claimed by evolutionists, why are there any creatures on earth or ever have been in that they would have been evolving backward from the very start, not forward. To succeed, this process had to be locked in for one direction or the other. At what time did the change take place? Then, as the evolutionists will argue, creation has a fundamentally irrational design, with an economy planned for absolesence. Why then are we still here? In the time frame they use, total extinction could have taken place millions of years ago. To argue whether a Zebra is a black animal with white stripes, or a white animal with black stripes is futile.

Doesn't this run counter to, and put a lot of strain on evolutionary adaptation? Dr. Gould says fossil invertebrates live five to ten million years and vertebrates a shorter time. In considering the Zebra theme why not say six to twelve millions years as long as we keep it 50% of our total and neither are provable?

If chance mutations or accidental choice made all of the life forms and kinds we now have, why shouldn't we be able to cross genetic boundaries much more readily by conscience selection (deliberate, intentional, not by chance or natural)? Yet, this is not the case. In animal husbandry it is possible, through conscious selection, to breed animals to favor certain qualities. If chance mutations kept the stripes on a Zebra no matter which part of the earth the animal may be taken to live and breed, would it be possible through a long period of conscious selection to breed out the stripes? Could we select black or white? If this could be done, it is no accident. On an earth wide basis why should regional adaptation need to accidentally decide between white or black for Zebra stripes?

If the Zebra were a solid color would it still be a Zebra? If a Zebra could be bred to black or white through conscious selection, would a genetic boundary have been crossed? Obviously not. It seems direct creation made the Zebra as it is. Who is able then, to fool mother nature; chance, accidental, natural selection, or conscience selection? For the evolutionist to say they have the answer to our puzzle will only mire them deeper into trouble. If conscious selection could not take the stripes off a Zebra, how could more radical change take place by accidental chance?

MISSING LINKS

In any theory we must have accountability from inception to terminus. Hybridization is not speciation. It does not perpetuate. If at random over millions of times a cell division makes a genetic copying mistake (a mutation), why do we get a weaker deformed life kind? If natural selection works at all, it works against itself, as natural selection discards the mutation from the species. They are harmful. There is a very good article in the April, 1982, issue of Life Magazine by Francis Hitching. The heading is: "Was Darwin Wrong?" Here is a short paragraph from this fine article: *"Yet, for all its acceptance as the great unifying principle of biology, Darwinism, after a century and a quarter, is in a surprising amount of trouble."* Another quote from same: *"Darwinism (or neo-Darwinism, its modern version), on the other hand, is a theory that seeks to explain evolution. It has not, contrary to general belief, and despite very great efforts, been proved.''* This article goes on to say, *"Darwin, like every other evolutionist before and since, turned repeatedly to the fossils for explanatory evidence . . . The minor 'improvements' in successive generations should be as readily preserved as the species themselves. But this is hardly ever the case. In fact, the opposite holds true, as Darwin himself complained: 'Innumerable transitional forms must have existed, but why do we not find them embedded in countless numbers in the crust of the earth?''*

Yet, the fossil record shows fully formed creatures appear out of nowhere mysteriously, with illogical gaps before and after them. Even more maddening, they show little change during their existence, then abruptly disappear without rhyme or reason, according to Darwinian evolution. Paleontologists and evolutions must now try to jump the fossil gaps. Can they succeed in replacing gradualism, with what they call "punctuated equilibria"? Will Darwinism survive the death throes it is in? They have fooled millions of people before, perhaps they have been underestimated and will again. Throughout human history, "truth" has never been popular. I would say, there should be gaps at the end of the Cambrian, Devonian, Permian, Triassic, Cretaceous, and other periods. The annular ring systems discussed in this book, deposited them in installments to the earth. The fossil record is just as it should be because of this unavoidable truth. Read this article I mention in the April, 1982, Life, and you will agree with my theme. Then, perhaps too many paychecks depend on the status quo, of saving face at any cost.

Endless new discoveries come in supporting the annular theory. (Ap) press release dated March 21, 1982 reveals a trove of animal fossils found in Antarctica, include first bones of a land mammal ever found there. National Science Foundation said the mammalian fossil find *"ranks as one of the most significant scientific discoveries in recent years.''* Other animal bones were found there, rare land lizard, sea reptiles, and man-sized penquins. This all proves the earth was once tropical from pole to pole, in support of annular theory. (AP) Oct. 15, 1982, Mud Butte, South Dakota — sixth discovery in the world of a tyrannosaurus dinosaur. It was identified by Phil Bjork, director of geology of the South Dakota School of Mines and Technology, Rapid City. The South Dakota fossil find indicates a cataclysmic destruction of short term. The tyrannosaurus fossil was found exposed on a side hill. If we can accept the erosion figures of geologists the fossil could not have been lying there for millions of years. The antarctic fossil finds indicate there was a much warmer climate there at one time than is now. Also, that there must have been a land bridge, or nearly so between the southern tip of South America and Antarctica when ocean levels were much lower than they now are. There are islands between Cape Horn and Antarctica. Life migration across to the antarctic would be possible with much lower ocean levels before the great augmentation took place.

As long as we are on mathematics, if we flip a coin once our chances are 50% either way. In the first few turns we have a lot of deviation either way, by chance. If we flip the coin one billion times we would have a total closure with no chance of gain on either side. So it is with evolution, time and chance destroy the theory. In fact, both time and chance completely wash out the very leg evolution stands on. A quote from Newsweek: *"If Darwin dealt a blow to the notion that man was created by divine edict, Gould has gone further and upset the marginally comforting image of evolution as a gradual climb up the slope of perfection, at whose crest stands man."* To say man is not special in many ways, is ludicrous. Yes, I will say this as a scientific point of view if you prefer.

The doctrine of evolution has about reached its apex for our time. It has been lying dormant until recent years when a new young group of paleontologists and scientists seeking prominence revived it. If the issue had not been pushed so hard, it would have lived longer. Evolution will be forced out of its long standing entrenchment by publicity fostered on it. Publicity will focus a spotlight on it and it must begin to defend itself in this new light. Both sides must come up with their best arguments and this will expose any weakness in both. Dormancy was a good hiding place for evolution.

The spotlight was on the bitter fight over evolution at the Chicago Conference on Evolution in 1980. After this bitter fight on the subject, Gould said, *"Evolution is a fact, like apples falling out of trees. Darwin proposed a theory, natural selection, to explain that fact."* Isn't it a fact, most of the theories proposed by science from Aristotle (384-322 B.C.) and Ptolemy, and even today, are not correct and the reverse is often true? It is implied that lack of education and ignorance has caused people not to accept and agree with evolution and other unprovable claims. We would assume that the highest educated people attended the Chicago Conference on Evolution. Now, if this is the result of education, total disagreement among themselves on the subject, imagine the quarrel if everyone were taught the same things and brought into the quarrel. Perhaps there could be revolution just as it has been among the great religions of the world for centuries.

After many years of study I see evolution losing ground constantly and continuously because of facts against it. In the quarrel they have among themselves, we hear such words as heresay, heretic, and apostasy, applied to those who do not go along with all of the new theories such as genetic gap jumping and others. If the Christians had not been persecuted they would have survived as they did with persecution, and those doing the persecuting would not have been exposed and embarrassed. The pressure the church put on Galileo helped his cause and exposed the church and those scientists who disagreed. So it is with evolution today. It is good that the issue is being forced by them as this will bring it out of the woods where the public and pedestrian can see and decide for themselves based on facts. Evolution has about reached its apex in our time and will decline sharply in acceptance. Error cannot perpetuate in the presence of light and truth. The more it is forced the shorter will be its lifetime. Opposition from the creationist will not cause its demise; it is not a provable concept.

The question may be asked, has modern technology left room for new revolutionary change in scientific accomplishment? Yes! This book will change completely the old accepted processes of geology and world building theory. Annular process is a necessity in developing worlds from their primitive state. The igneous fluidity of the primitive earth set the mode of ring-formation above the primitive earth. Oil, coal, and carbon, were distilled in the fiery fluid earth; and, along with water and other elements, were thrown into rings around itself. As the rings fell back to its surface, they built the coal and oil beds and its crust just as we see it today. Proving, as this theme has, that carbons are not made of vegetation is a total revolutionary departure from current accepted geologic philosophy. As one reader from Illinois wrote, after reading my first book, *"Your book seems too sound to be called a theory. You have demolished the idea of vegetarian origin of oil, etc."*

PREDICTIONS, PAST AND PRESENT

The issue of life existing on other planets and, for that matter, the Universe, is largely a subject of Mythology. Most of us are familiar with the ramifications of Greek Mythology and its affect on the modern world. Many things, including planets, are named after a Greek God or legend. Evolutionists and scientists scorn doom's day predictions by religionists and others, yet they do the same thing on a larger scale and justify themselves because of the intellectual source. When their predictions fail, which they do most of the time, it was just an educated error — it is excusable without reflection. Is this really sound? No, it would seem not.

We may reflect on the doom's day predictions of astronomers for the date of around March 10, 1982 when the planets would be lined up more than usual. Great cataclysmic disasters were predicted by astronomers ten years in advance of this date. May I proudly say I disagreed with them from the very outset. It has always been my stand that the planets were never programmed to align themselves to self-destruct. I have several books predicated on this assumption. One book first published in 1963, and still available, says: *"The trigger for cataclysm has been set, and is from an outside force."* This book goes on to say, *"On May the 5, of the year 2000 A.D. the planets of the Solar system will be arrayed in practically a straight line across space, and our planet will be subjected to enough gravitational distortion to tip the delicate balance."* The author says, *"This lineup of the planets will cause centrifugal aberrations which will change the earth's rotation and equilibrium."* The road to astronomical doom's day, seems to be very crooked and uncertain. We might ask ourselves, why should such planetary configurations have such a devastating affect on the solar system? Are there other conjoining factors we may have missed? I know of none. Then if the alignment of March, 1982, did not cause disaster why should May 5, 2000 be any different?

"Close Proximity of Mars May Show Planet Inhabited": The question of whether or not the planets support life has long afforded the scientific world an interesting subject for conjecture and speculation. Many weird tales have been written about the strange race of people popularly supposed to inhabit Mars." The above quote could easily be right out of Scientific America, Discover, magazines or many other popular scientific publications. No, this quote is not from a current publication. It is the title and first few words of an article published in The Pathfinder Magazine of August 26, 1922. Now that this educated scientific prediction did not come true they have just moved it farther out in the Universe. Reflect on this for a moment. It is not unique and the same failure of prediction will be manifest.

At that time, many disagreed saying the condition on our celestial neighbors do not indicate the existence of any kind of life. Yet, astronomers said everything is favorable to both vegetable and animal life on Mars and Venus. Huge telescopes at that time were being built around the world and they would prove life existed on Mars and Venus. David Todd, American astronomer, and B. McAfee, another noted scientist, felt the luminosity of the new telescope apparatus would show Mars supports life. Haven't we heard all of this before? Both Professor Lowell of Harvard University, and Flammarion, consistently held that Mars was inhabited. They felt, because at times strange sounds heard over high-powered radio receiving sets were too regular to be the work of nature, these so-called "mystery messages" were the work of living creatures. Many, however, pooh-pooh the idea. Aren't we still hearing these stories from modern astronomers? Yet, some scientist back then believed these signals were messages being flashed from one or other of the planets in an attempt to open communication with the earth. Sound familiar!

The last suggestion was that the 47th proposition of Euclid (Greek Mathematician, 300 B.C.), concerning the three sides of a right triangle, be outlined by millions of electric lights stretching over several states. It was argued that if this were done, it would serve not only to signal the planet, but to test the intelligence of its inhabitants as well.

It was believed then, that the only way to establish communication with Martians would be by a signal or radio wavelength. The idea of sending a projectile through space was considered. The Jules Verne (French novelist, 1828-1905) story, a "Trip to the Moon", was felt to be absurd; it just could not be done. It was believed then, that even if a projectile or bullet was fired out from the earth, it would never reach its destination as it would be caught by gravitation and be held prisoner by the earth. Then, if it did leave the earth, it would not land on Mars but be caught by the gravitation of Mars and revolve about it as a satellite. There was some real logic in some of this theory. As with Verne, and others, they did not confine themselves to scientific facts; they made assumptions that would not hold good in practice. As the article says: They wrote plausibly, but that is all. Isn't this exactly how astronomers and scientists are today? There was a man back then who was working on a plan to have himself shot to the moon on a rocket; it was felt to not be sensible then.

They reasoned then, that a projectile (now a space capsule) traveling with such speed as to enable it to clear the earth and not fall — would be heated to incandescence by air friction, and would be consumed, just as shooting or falling stars are. Space flight, as we know it, was not understood back then, though they had a good understanding of the problems. They felt no sensible person should risk any money on a project to shoot a man to the moon on a rocket.

Next the astronomers, as this article reveals, believed the so-called "canals" of Mars are regular and straight lines which seem artificial — the deliberate work of Martians. They reasoned that rivers are never in straight lines so these canals must be man or martian-made. Yet, many authorities said the artificial appearance was deceiving; a freak of nature or an optical illusion. Then, as now, it is difficult to change the minds of astronomers and scientists, even with facts. They even felt the glistening white discerned on Mars proved snow and ice at the polar extremities. Areas of green, often visible on Mars, was thought to be vegetation or perhaps seas. Scientists back then who were well informed, agreed that the atmosphere on Mars would be so rare that living beings could not exist. If they did exist, they must be quite different from anything we know on earth. Professor Lowell of Harvard University once said, *"I have absolute proof that Mars is inhabited. Every observation I make increases my conviction that Mars supports life. Nothing contradicting my theory has ever been discovered; and, on the other hand, confirmatory evidence is constantly being brought forth. The smaller mass of Mars should theoretically make its evolution somewhat different from that on earth; its surface must be smaller and more level and its air scantier. Observations establish precisely these facts. . . . Consideration of temperature indicates a climate of which the warm season, the one most favorable to life, is hospitable to both kinds of life, vegetable life revealing its presence by seasonal changes of tint. Then the consideration of water supply shows it to be scanty but present, as evidenced by the polar caps which must be snow and ice."*

It seems the reasoning and theories of modern astronomers and scientists is not far removed from that which was expounded at the turn of the century. Instead of denying the accuracy of a hypothesis they will just expand it to cover more ground. Again, much of current geophysical theory must be discarded totally, just as the Ptolemy theory has been. Modern astronomy has much of the old theory blown out of proportion, such as the Cosmos TV Programs. Going back to 1900-1922, the work of Professor Slipher (Lowell Observatory), Dr. Asaph Hall, Professor Lowell, and their contemporaries we see a trend analogous to the modern astronomer and paleontologist. To analyze: Dr. Asaph Hall, one of America's foremost astronomers of the time I mention, expressed the opinion that there must be a little atmosphere and less heat and light on Mars. Inhabitants, if any, must be quite different human beings as we know them to be.

After intimate observations by astronomers of that time through the telescopes they had, no remarkable discoveries were recorded on Mars or other planets. However, Professor Lowell expressed the opinion that the Martians are long limbed creatures, with grotesque heads, and arms that hang almost to their feet. Astronomers continued the assertion that even a cursory inspection would prove the threadwork of canals on Mars were unnatural in regularity, and that this adherence to a definite system of traits of their artificiality. They suggested the canals cannot be something of a natural origin; therefore, the necessity of intelligence to develop them was evident. They were certain the canals were built by Martians to bring water from the poles to where it was needed. Doesn't this appear to be exactly like our very own Bureau Of Reclamation, Army Corps Of Engineers, and Department in Interior? Perhaps they already had bureaucrats on Mars.

In these reflections you will see that I am making a very helpful point for our decisions on this subject. The astronomers then, as well as now, were a very obstinate group of men. Certainly much has been learned from their mistakes and wrong conclusions. They were wrong more than they were right, just as it is today with such postulation. There is an obsession and enchantment for some scientists to prove life exists somewhere out there in space. This obsession sometimes clouds their minds from plausibility. For example, they reasoned that Mars was older than the earth; therefore, any life on an aged planet must be of a higher order. We would assume they mean it had more time to develop and evolve. This life of a higher order, as we know, turned out to be nothing at all, not even a bug or microbe.

In order to reflect back, and also forward, I have read all of the information astronomers and scientists have been able to come up with; yet, I see no proof or reason to believe there is any form of physical life away from the earth. Tens of thousands of pages have been written to show it would be possible for life to be out there somewhere, none of it plausible. On the other hand, a lot has been written to show why life does not exist away from the earth, and of all of the plausible reasons life doesn't exist in outer space, only one is needed. It is a substance called ''oxygen''. Just as genetic speciation does not cross a genetic boundary, perhaps in the Master plan of things, life has not spread to outer space and other planets, not until now at least.

An interesting discovery about oxygen is the high concentration of oxygen in antarctic waters, wherein one species of fish has no hemoglobin in its blood. This white-blooded fish lives without red corpuscles; it would die in another environment. It seems all life found so far cannot live in a vacuum, or without oxygen.

There has been endless material published on the subject of colonizing other planets and outer space. It only stands to reason if this theory were carried out on a grand scale, as some suggest, and, things necessary to sustain life out there had to be taken from the earth, the time would come when there would not be enough life sustaining substance left on the earth for those of us who wanted to stay. Perhaps in a fictional way we would have the basis for a universal war between the two factions, terrestrial and spacial. Maybe a nonsense TV Program could be built around this subject. About every other known subject has been spent on TV and they have nowhere to turn.

In recent years, there has been almost a desperation to prove life exists on other planets. Many fiction TV shows about space, etc., were linked to this theory. There was a big fuss made when the first moon rock was brought to earth, so as not to contaminate the earth with some new form of bacteria or whatever. Now they are manufacturing them (gene splicing). I have always believed the other planets were sterile of any life, not cell, bacteria, or microbe. I hope science does not invent proof of life on other planets if there is none. They have been inventing materials for years on this subject. If I am proven wrong about life in outer space, I will not be ashamed rather happy to know the truth.

It seems ironical that astronomers and scientists would spend so much time searching for life in outer space when they do not yet know how life got on the earth and how the earth developed. It is sheer folly to talk of colonizing undiscovered planets. Here is what one astronomer says, *"If the human race survives the next 50 to 100 years,"* predicts astronomer Smith, *"we will have spread out into the solar system. Over the next 10 million years after that, we will populate the entire Milky Way, a galaxy that is 100,000 light years across."* U.S. News & World Report, Nov. 16, 1981. Sounds like Pathfinder Magazine, August, 1922! Senator William Proxmire of Wisconsin said, "There is not a scintilla of evidence that intelligent life exists beyond our solar system." U.S. News, same issue.

A REVIEW OF INFORMATION

After examining this wide variety of information, we will come to some definite conclusions. Oil, coal, gas and carbons are not made of organic animal and plant life. The oceans of the world have been augmented throughout geological times. The earth's vast mineral beds have fallen into place stratifying the earth from the bottom up to surface. No matter what record or source of information we may go to, every fact points to a spontaneous and direct creation of each life form and kind. The fossil record shows each life kind had no ancestors, forefathers, descendants, or relatives different from themselves in their family. After many years of research and effort to prove contrary, this is a fact and stands as such to now.

How did life and fossils get buried as deep as 15,000 feet in the Cambrian rocks? The trilobite, even from Paleozoic rocks, is not related to the modern day crab. It is obvious the lower forms of life were placed with an age when it was developed and each age was once open to the atmosphere. Could life have been placed in the ring systems as they developed by God, or whoever or whatever you may think? May I ask a question? Could it be the lower forms of life are not counted as part of the higher forms of life in the six day creation period, which still could be thousands of years in duration?

It is also reasonable that the earth once had its own ring system such as Saturn has today. That mineral and elemental segregation was done before placement not after. This ring system had all of the elements the earth now has in its crust and are found just where this ring system deposited them. Ocean shorelines have always moved up not down, and continents and mountains were always heaved higher, accounting for marine fossils high upon them. That the earth was once tropical from pole to pole. Plant and animal fossils prove this beyond any doubt. Ice caps were placed on the poles in a down rush of water and snow and are still right where they were placed. There is no replacement factor. A once living tropical and semi-tropical world still lies under them.

Vast metamorphic elemental changes did not happen. Fossils found in coal and other material are a foreign material in each, mineralized by their surroundings. Which did they make; coal, clay or rock? Vegetarian geology just is not a true concept, the process is too puny to create all of the carbons of the earth. Oil is now found thousands of feet below fossilization. Carbon is on other planets without organic life.

That hydrostatic compression of the oceans of the world by constant augmentation would not allow for continents to drift to any great amount. Hydrostatic compression would force continents higher not lower. The arch of a continent would always remain as such. Sea floors could be pressed down some with continents thrust higher as a displacement factor.

The material which forms the rings in circumplanetary development arose from the planets themselves and not from an outside source. This has been discussed before. The ring systems once on the earth arose from the earth and returned to the earth. If the ring material on other planets did not originate from the planet where would it get the independent rotation inertia required to sustain it? It is true, asteroids and meteorites may fall into a planet; however, this material would not become a ring system as it would lack rotation force. It would depend on which direction asteroidal material entered a circumplanetary system. If an influx of asteroidal material entered a ring system against rotation, there would be energy diminution due to collision causing material to descend to the planet or slow it down. If asteroidal material entered too fast and with rotation it may give a small tug to the rotation and curve on by and out the other side. If it entered too slow it would descend to the planet, just as a meteorite does today. Perhaps it is possible for some asteroidal material to join a ring system. Where would it get rotation inertia? We can say confidently such material did not form the vast varied rings on planets. Any large amount of material entering a ring system would have to borrow energy from the system, reducing its perpetuity.

If a circumplanetary system captured outside material, would this material give it a tug, or would it be diminutive to the entire system? All material matter is energy but this is not what we are considering here. If there were not a pre-existing ring system on a planet, why should outside material suddenly make up one? This cannot be the case. We must assume that any action interfering with a ring system would reduce its energy and the entire system must fall back to the parent planet. It is true, no matter how much we may theorize and move away from home base, we always end up right where we started from. We cannot invert or increase a fixed energy unit. It has been tried, such as to invent perpetual motion. We are back to the story of a ball rolling down a flight of stairs. It will slowly come to the bottom and remains there as in any action, it is seeking neutrality. So it is with a planetary ring system. Broad sweeping claims will do us no good. Dr. Carl Sagan said, *"He could not predict what would be 50 years from now, but could predict what would be several billion years from now."* (TV)

It is my turn to make a prediction and it will be that massive radio telescopes will discover more mysteries which cannot be answered. Man cannot answer all cosmological questions. Perhaps every atom may be a universe. If so, we are on the edge of forever. Some scientists seem to lean toward the Hindu Religion as it lends itself to the evolving theory of the universe. A good question is what is a religious fanatic? We must first decide what is a religion. Was Christ a religious fanatic? Is the Pope a religious fanatic? Is Billy Graham a religious fanatic? What about Martin Luther or Mahatma Gandi? Would you say William Tyndale (Bible translator, 1494-1536, executed for heresy) and other translators were religious fanatics? The doctrine of evolution has become a religion in our time. Would this make Dr. Sagan a religious fanatic, or the Billy Graham of evolution? If the evolutionists could get us all to accept their ideas and theories and abandon all others, and follow them, wouldn't this be a very large religion in need of a god they are so sure does not exist? Evolution should be a science, though it has become a very large religion. Anything worshiped is a religion. Evolution is a religion in that it can't be backed by observable facts. We may ask, what is the cornerstone of evolution, if there is one? Perhaps flexibility is their main cornerstone. A simple question never answered by evolution though accepted by most persons, that dinosaurs were both plant and meat eating animals. How do they prove this? Perhaps they were all plant eating. We are forced to the conclusion there can be only one source of religious Truth, the Scriptures.

See if you can make sense out of ths one. A geneticist was on national television in support of gene splicing and recombination. When asked how he could be sure they would not create by accident and mistake some unwanted and dangerous thing. He said, in their experiments there may be only one mistake or accident in millions, and this would not make any difference. The only assurance he could give this would not happen, were maybe, we guess, we don't know for sure, what will will result. What, then, about all of the drugs being used that cause deformities and death? We all know what man-made drugs and compounds have done. How can we be sure they will not do even worse with their new money-making scheme and plaything, gene splicing?

Then, on the other hand, technical science has been telling us for years that accidental and chance mutations, are the building blocks evolution used to develop all of the beautiful life forms on earth. Now that they want a free hand in this new scheme of gene splicing, why are they assuring us accidents will not happen, or in such remote possibilities it will not make any difference. If chance and accidental mutations built all of the life forms on earth from a single cell, by evolution, as they claim, why are they now assuring us a few more brought on by mechanical means will not make any difference and may not happen? Why do they now say will not happen what they formerly said must happen? They are admitting their mechanical choice cannot do as well as blind accident. Can you make any sense out of this inconsistency and complete turn around? Another 25 years of technical science as the last 25 have been will leave the earth a sterile ball, unlivable.

A HETEROGENEOUS WORLD, WHY?

In the progression of a heterogeneous world the question may be asked, if centrifugal force played such a prominent role in building ring systems, why didn't all of each element form a single solid ring and all fall as a single ring? First, the ascention and declination (rise and fall of rings) of annular material must complement each other in order and reverse order on return fall. Rings on a planet are carbon copies of each other in declination, they must fall as they arose, in order of segregation and assortment. Why then, didn't all of the carbon rise together and fall together? If we knew how the original globe developed and was placed where it is we would know the answer, but we do not know. Because oil, coal, and gas, are found from earth's surface down to 30,000 feet deep we know this did not happen. Likewise, water is found deep in the earth. Each then must have arose from the igneous earth in installments and returned in the same order, as did other minerals and elements.

The densification of volume and distribution of gold will vary in the earth's strata down to 8,000 feet deep. Gold may be found on Alaskan Glaciers and all over the earth in its crust. No one knows how the original mineral and elemental distribution developed in the original globe and other planets. Because these materials were delivered back to the earth in varied installments, they must have arose in the same way. There would be some water in most of the rings with the greater amount segregated to the outer rings just as it is on Saturn. Both known and unknown forces did this assortment work.

Carbon had to be a very prominent part of the original earth and other planets as it is still there and here. Carbon must have been the legitimate energy which built the planets and the Sun. Perhaps it was the igniter which brought other materials to flame. It is still a universl product and energy source. When we examine the world upon which we live, with an impartial and philosophic view, we see stupendous revolutions are chronicled in its crust and strata. It is overpowering in magnitude. Carbon must have been the igniter of all other materials and carried the flame.

The igneous primeval earth was composed of a mixed fund of silicon, calcium, iron, copper, lead, silver, gold, sodium, oxygen, hydrogen, and an immensity of carbon. Carbonates were there; so the measureless fund of carbon now stored away as coal, in amounts beyond all comprehension. We may not know every detail; yet, we see the great universal process of development. Technical science has found everything right where it should be. Many discoveries may be a result of serendipity. Many fossil finds are just accidental finds by local people or someone looking for something else. I do not mean to imply that gravitational and centrifugal law have ever been in violation of each other. There must be a great outside force applied, in this case it was inveterate heat. As this great heat dissipated, an adjustment took place in circularity of centrifugal law in relation to gravital. It stratified the earth — we see the results before our very eyes. Law demands that the earth must be heterogeneous and that is just what it is. The earth would not support life if it were not. Both religious and scientific persons claim the earth will burn up and be destroyed. To the religious person may I say, the scripture says the earth (not world) will abide forever. To the scientific person to destroy the earth would throw the solar system out of gravital balance and all planets into disarray and end the solar system. Neither will happen which I hope will not upset anyone.

Our agency for a heterogeneous world was igneous, centrifugal, gravital, and cataclysmic upheavals and some blending. Metamorphism had little, if anything, to do with the elemental development of the earth's contents. After placement when elements are removed from an igneous agency, no metamorphism is possible.

CONCLUDING THOUGHT

May I say I am not satisfied with my effort in writing this thesis. Perhaps no writer is ever satisfied with his work. It seems I may have repeated too much. Or perhaps left out some point I should not have. Or perhaps could have made some points clearer than I did. To write on such a fantastic subject, with years of study, it is necessary to condense. The problem is, how much should be condensed out, and still bring an accurate chain of thought to a logical conclusion. It seems to me a 500 page book would be more in order. Yet, I hope I have been able to be helpful to my reader and bring a much wider view of this subject to his mind. It was not my intention to be contrary to orthodoxy, yet if this is the only way a subject may be presented, so be it. Thanks for reading my subject, and only wish I could have a talk with you in person, though that may not be possible. It has been my privilege to meet so many interesting people in my travels and work. This has enriched my life.

When the mind is staggered by the reality of facts before us, we sometimes close it as a defense against this reality. Should we allow prejudice and preconceived ideas to set mental limitations for us? Mental oppression is sometimes self-inflicted. Why is the last recourse "truth"? Because orthodoxy has taught us many things in reverse, are we to abdicate to error out of a seemingly false necessity? Let us smash through these mental barriers and learn facts and stand as we should in this light. Should intimidation govern our minds? Truth and light are the only basis for judgment. Peer pressure and propaganda are powerful forces in mental development and decisions. The "world" around us wants us to believe something because the majority believe. You, my reader, I am sure, are above this kind of mental intimidation.

ACKNOWLEDGMENTS

May I say how much I appreciate all of the information of the people, and scientists who have contributed so much to the information on creation and the origin of life. I have mentioned many books, magazines and authors, in the main body of this thesis. There are so many hundreds more I cannot list them all. This does not mean I am ignoring them, and their contribution to the subject here discussed. It has been with gratitude I have examined much thought from intelligent authors and researchers. We must always examine every pro or con, on a subject, to draw a sound conclusion.

Do not get the impression I am suggesting no one should read the books and articles I may not agree with. I recommend highly all of the material I have referred to. It is all good reading, and very educational, and informative. The Fossil World is very educational, though I disagree with the way it is applied. All of the geological material, and others, I refer to are very valuable and informative to read. All of the news magazines are good, including Reader's Digest, Smithsonian, and National Geographic, and others too numerous to mention. We must provide the intellect material to build on.

A special thanks to Jan Froehlich for typing my manuscripts. This was a great help to me in that I do not like typing. Also Rod Stoa, Curator of the Wilson M. Laird Core and Sample Library, Grand Forks, for his help with pictures and information. North Dakota has a very excellent and complete core library. Thanks to North Dakota Geological Survey for their help in gathering information. A special thanks to the Foster County Independent newspaper and all other papers and oil journals for news and publicity about my first edition book. Special thanks to Mavis A. Vogel (Editor), for her scholarly full page article with pictures in the Independent on my first book. I must include the many television stations who gave me time for interviews about my geological theme. Channel eleven at Fargo gave me one-half hour and ran it several times.

Appreciation to Tenneco Inc., Amerada Hess Corporation, The Petrified Forest Museum and Park, of Arizona, and others for such beautiful pictures. The American Petroleum Institute, for geological structure diagram. Many thanks to the authors and scholars who supplied me information over the years to add to this great store of information I gladly share with you. There are so many I cannot name all here. I must mention again Isaac N. Vail, who contributed so much to geological wisdom and world creating processes.

It is with deep respect, and credit to Isaac N. Vail, that I here have included the Preface to his book, "The Earth's Annular System."

PREFACE

On the 26th of January, 1912, Isaac N. Vail, the author of this book and the originator of the Annular Theory of Evolution, died suddenly at his home at 411 Kensington Place, Pasadena, California.

About the first of December, 1911, the previous edition having been exhausted, he had made arrangements for a new edition of this, his largest published work. Only about half the plates were finished at the time of his death, so that it became the duty of his daughters, Alice Vail Holloway and Lydia C. Vail, in whose possession he left all his published and unpublished writings, to carry on the work. This volume now goes before the public just as the author left it.

Isaac N. Vail was essentially a student, modest and retiring in his nature, and with but little of the aggressiveness of either the propagandist or the man of business that would have pushed his views before the world. In his later years, he found increasing contentment in searching out material to be used by those who would continue the work after he had left it.

No one realized so fully as he the far-reaching results and revolutionary effects of annular thought upon almost all departments of science and philosophy. With this realization, and with a confidence and patience born of a knowledge of the truth, he was willing to wait till the world was better prepared to accept his ideas.

He knew that in the light of new discoveries, old theories must fail, and that in time, the Annular Theory of Evolution must gain a sure foothold in the young and vigorous minds of coming generations.

To the end of carrying toward completion, the work begun by our father, our lives will be devoted, and as the demand arises, we hope to give to the public further evidence in support of annular evolution.

We undertake this work, not only with the feeling that we are fulfilling a sacred trust, but with a love of the work for itself, and with a profound conviction of its importance to the progress of science.

While we may be able, from time to time, to present to the world even vast accumulations of evidence gleaned from the fields of geology and especially mythology, we can do but little more than suggest to more able and scholarly minds than our own, the work to be done in the fields of Philology and Biology.

A new science of the origin and growth of language will have to be written. The Darwinian theory of evolution will be found insufficient and will have to be supplemented by the Annular Theory of organic evolution. Following this must come a new Ethnology and a modified Sociology.

It is with the hope that we may gain the attention of those who are able to do better work than we, that we now send forth this volume.

ALICE VAIL HOLLOWAY.

LYDIA C. VAIL.

Pasadena, Cal., April, 1912.

It gives me a great deal of pleasure, and satisfaction, to be one who in our time, to bring to this generation, additional facts and truth, which will uphold factual world-creating geological process, accounting for the tremendous variation in the carbon, mineral, and life forms so overwhelmingly demanding explanation. As scientific as this all may be, it upholds the Biblical account of creation. For those who do not believe in a creation by a CREATOR, nothing will change and they too will enjoy the information and benefit from it. All through historical times, great men have stumbled over simple truth and fact. Start with Copernicus, to Galileo, to modern times, and include if you will, Vegetarian Geology.

May I say, overcoming the statistical odds of evolution are staggering. Sheer intractability of the problem causes skepticism. Professor Vail would have been very happy to have heard these words by Ernst Chain, who won a Nobel Prize for research into the curative properties of penicillin. He said, in 1970:

"To postulate that the development and survival of the fittest is entirely a consequence of chance mutations seems to me a hypothesis based on no evidence and irreconcilable with the facts. These classical evolutionary theories are a gross oversimplification of an immensely complex and intricate mass of facts, and it amazes me that they are swallowed so uncritically and readily, and for such a long time, by so many scientists without a murmur of protest." (LIFE, April, 1982)

"If it could be demonstrated that any complex organ existed which could not possibly have been formed by numerous successive slight modification, my theory would absolutely break down." — Charles Darwin, "The Origin of Species".

Even in Darwin's time there were many natural wonders that made him very nervous about his theory. With the information now available he would have been sleepless. For those who reject direct creation, no one knows how or when the genetic code originated. To now apply the quantum jump, as science has, to the complex genetic fixation, would frighten Mr. Darwin. He once said, *"The eye makes me shudder."* To say that competition among the species intensified natural selection, when competition was nonexistent, is stretching the point too far in desperation.

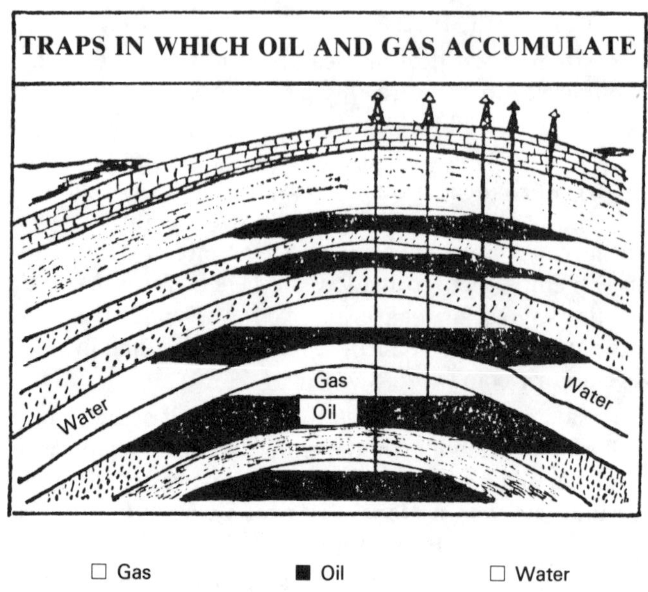

☐ Gas   ■ Oil   ☐ Water

It is very hard to draw an exact picture of how the oil bearing formations of the earth really are as no one has ever been down there to look at them. This end view diagram will give you an idea what an anticline may be like. The great Nesson Anticline in the Williston Basin may appear like this. It is from a mile wide to perhaps several miles in places and is more than 80 miles long. Oil, gas, and water, are never found in great open pools or lakes in the earth. They are always found within the boundary of rock, shale, limestone, or whatever they are in. Some places may only have one oil layer, where others may have more, even many. Some places in the Williston Basin oil may be found in four, five, even six levels deep as 16,000 feet. Each layer may have a different oil and specific gravity. On an average the U.S. produces about five barrels of water for each barrel of oil. Dead marine life could not have made many layers of different kinds of oil and gas, in many different kinds of rock, with different bottom hole pressures, and different kinds of waters. Some bottom hole temperatures are very high. It seems oil was distilled in the igneous earth before organic life, as were many other minerals and materials.

This is not to mean oil is only found in six geological zones or layers. There are geological reports which show as many as 32 different geological zones or ages produce oil or gas in the Williston Basin. Again, this would vary from well to well and area to area, and also on an earth-wide basis.

WHERE PETROLEUM IS FOUND

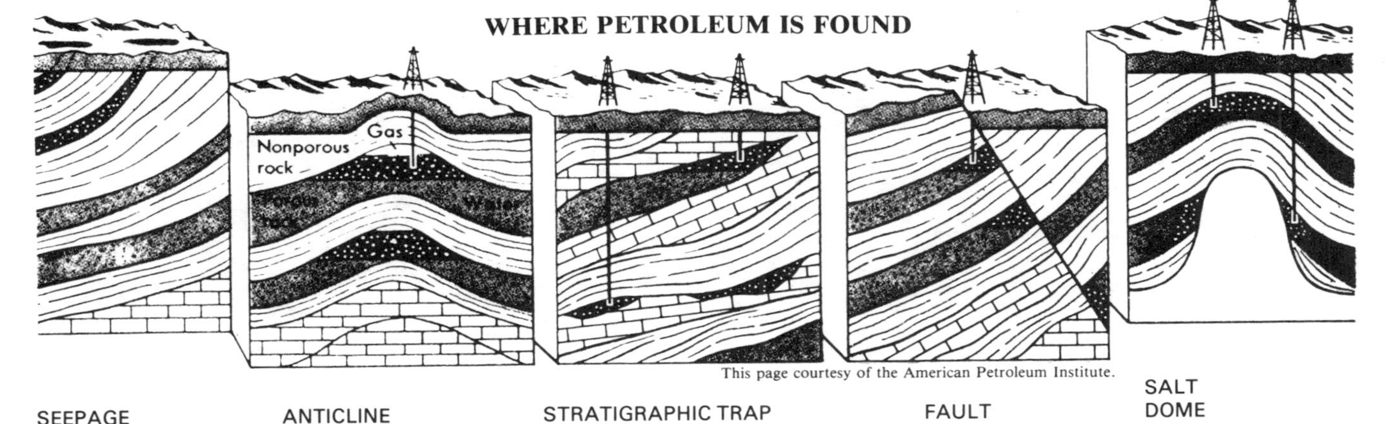

This page courtesy of the American Petroleum Institute.

Petroleum Collects in various kinds of underground structures called *traps.* These were formed by changes in the earth crust. Droplets of oil collected in the tiny spaces found in porous rocks. Layers of nonporous rocks sealed off the oil-bearing formations.

ONLY THE BIT PROVES THE PRESENCE OF OIL

In spite of years of study, the best that any instrument can do is to locate structures which may contain oil or gas. Persistent efforts have been, and are being made, for a device that will actually find petroleum itself. No scientific instrument showed the presence of even a potential oil reservoir in the great East Texas field, which is a stratigraphic trap. Other great fields of this type may be discovered elsewhere if enough wells are drilled. For this reason, the petroleum industry spends millions of dollars annually to drill wells in areas of questionable geologic promise. (*Research in locating oil by electronic means without drilling shows some promise.*)

Looking north over The Grand Canyon, notice the plateau across and the horizontal stratification, with little or no upheaval in this area, indicative of a fluvial cataclysm, not slow constancy of erosion.

This scene is out of Palm Springs, California. Notice the slant formation upheaved in appearance (Piercement) as are most mountains in the western U.S. and most of the world. Not erosion.

Geologists claim it took 1.7 billion years of erosion to make the needles eye, and other needles. This cannot be correct as erosion would only build flat lands and plateaus. The rocks may be older than 1.7 billion years; however, a cataclysmic deluge of water fell washing away the softer material and rock leaving the hard material as we see it here. Slow constant erosion could not create such irregular formations (South Dakota Black Hills).

The Great White Throne, Zion National Park, Utah. Said to be the largest monolith in the world, 2,447 feet above the canyon.

The Temple of Osiris, Bryce Canyon National Park, Utah.

In the accepted theory that slow erosion was the main energy factor which built the two structures above — the erosion must have covered wide reaches of the earth, relentlessly. These two (page 232) radically different structures are only a few miles apart. How was selectivity done over millions of years of constancy? One is solid rock material, the other is softer material. Millions of years of erosion would have left the Bryce Canyon scene horizontal. A single force over millions of years would affect each differently. It seems a great down-fall of water built them both, by washing away the looser material, in a short time span.

What then, built the great granite beds of the earth? Some may be found at high elevations and other granite is found thousands of feet underground. Why should erosion first build them, then place upper fill on top of them? The Williston Basin has a great granite underlay-ment, which outcrops far to the east in Minnesota. Can erosion per-form miracles?

How do we account for billions of tons of marble high up in mountains, and other places around the earth? By erosion? Marble, as well as granite, must have had an igneous agency (heat or fire) applied some time in its origin. Since the time of Augustus Caesar (B.C.) the quarries at Carrara, Italy, have produced an inexhaustible supply of marble for Rome and the world. This marble is in Emperor Trajan's Column and the Pantheon. There must be trillions of tons of marble on earth. Are we to believe that Coral and other marine organisms built the vast marble beds of the world, by becoming calcium car-bonate and limestone crystallines, under water? Could crystalline marine corpes become marble through millions of years of metamor-phism? We all know that the food supply for a creature must exist before the creature. In this questionable process the food supply becomes the marble through the food supply of the marine organism. If we feed a cow one hundred pounds of food, it will not make one hundred pounds of beef. Here, we are talking about trillions of tons of marble. Thermodynamics says we cannot make more out of less. In this postulated process how many tons of food and marine bodies would be required to make a ton of marble? What, where, and when, was this food supply made available? If a source could be found, wouldn't it fill a large part of the earth? To further compound this problem, science also has marine life building all of the oil and gas content of the earth, at about the same geological time period. We may ask, how do the fish know what to do, make oil or marble?

Then, isn't it true, organic material never becomes inorganic, such as marble. This accepted theory for the creation of marble agrees that an igneous agency was applied with pressure at one time. Okay then, at what time in the process could the igneous agency be applied, and not destroy the organic food supply of the marine life, or the life itself? Any theory that fails in one point fails completely.

Even if we could cross all of these hurdles, would we have the answer and would the problem be solved? If we said no a thousand times it would not be enough. The greatest hurdle is yet to be crossed. Organic life may draw sustenance from inorganic material via the photosynthetic motor. One authority says this process lay dormant for 170 million years to solidify the results, make marble. The photosyn-thetic motor has no reverse gears by which organic material may become marble, the most extreme we could think of. The same old problem now plagues them, the convertibility of organic substance in the presence of oxygen, to inorganic materials. It just never happens.

Does it appear possible that this scene could have been buried underground for 200 million years with 3,000 feet of rock and earth on top of it? Some geologists claim this. If we add another 3,000 feet of regional elevation it would make it even more impossible that the Colorado River cut the Grand Canyon. Experts say about one-quarter of an inch of soil erodes steeper slopes each year now, less than half that amount from gentle slopes, and still less from level land. Are there still tree and dinosaur fossils buried hundreds of feet down in the earth? It would seem not. The Petrified trees must have always been right on the surface.

It is presumptuous to say the photosynthetic process was reversed to make marble. Marble just was not made of marine and Coral life, anymore than diamond was. In this marble development theory, how do they propose to transfer millions of tons of marble from ocean bottom to mountain peak? The conversion of marine life to marble just does not seem possible. Isn't this just another case where science has things in reverse order? If we accept the theory offered by science, wouldn't we have to violate the laws of transfer, conversion, convergence, and thermodynamics? The annular theory I have presented answers these problems. Someone said: modern classroom education in the sciences will become pandemonium. Perhaps it already has!

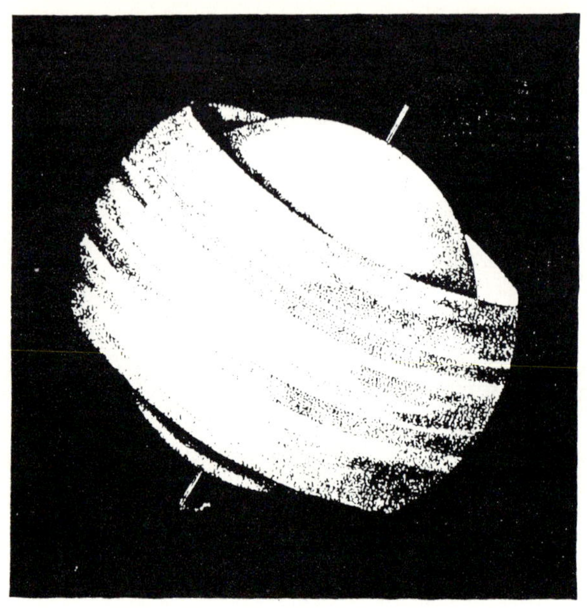

FIG. 12. EARTH IN EDENIC TIMES.
(CANOPUS AND POLAR OPENINGS.)

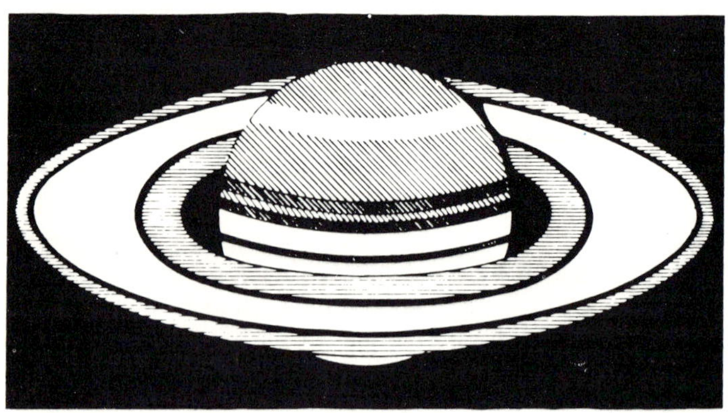

SATURN (RINGS FORMED)

Saturn is the crowning glory of the known Solar System. The rings are many, number unknown, but each one must eventually fall on the surface of the planet, close an old "age," and begin a new one, just as similar rings have done repeatedly for the earth. In Saturn's system we see the hiatus between ring and ring, and thus the vast gap and "missing link" between age and age, and life and life. (Isaac N. Vail developed this replica of Saturn over 80 years ago. It is just as the actual photos by Voyager 1.)

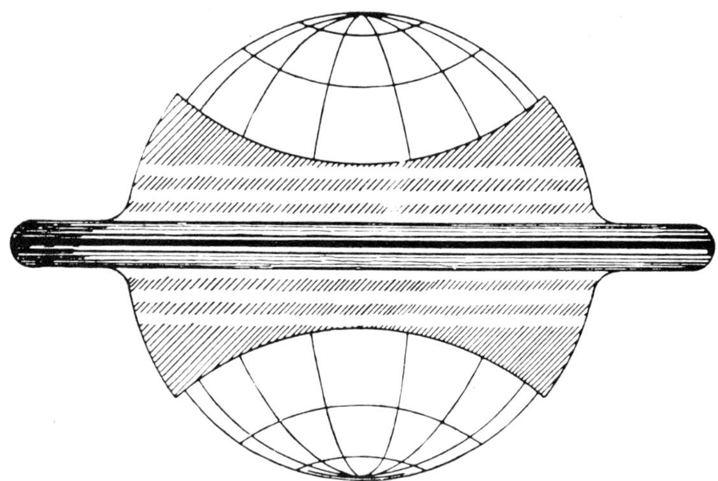

THE FIRST CANOPY SLOWLY SPREADING POLARWARD

This is an edge view of the Earth's Annular System with its innermost ring having reached the atmosphere in its slow and gradual descent spreading from the equator to the poles. Revolving rapidly around the earth, it is thrown into bands, belts and lines as it forms into a canopy such as the planets Saturn and Jupiter have today.

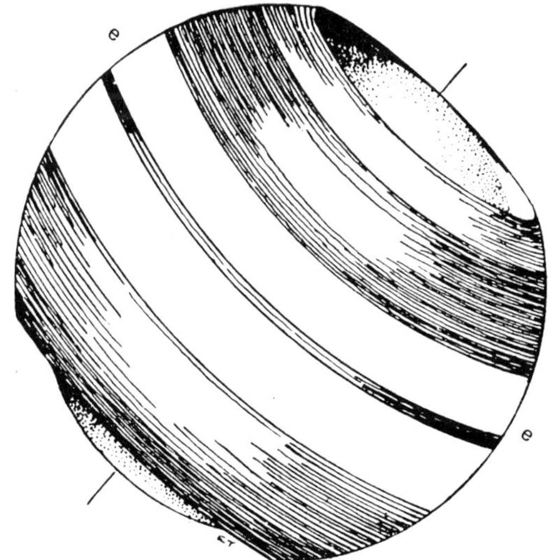

THE LAST CANOPY OF EARTH

The Last Earth Canopy must fall. It opens at the equator (e, e) and the vapors slowly float to the poles, and begin to fall. Snow begins to chill the earth. The sun shines in upon the equatorial earth through the opening and air in that region rises. This starts air currents from the poles and these currents bear the falling vapors back toward the equator, thus making one long-continued downpour of waters in medial latitudes. So, in the windup of canopy influences, there must be not only a vast accumulation of snows at the poles, but long-continued and devastating floods in warmer lands.

(237)

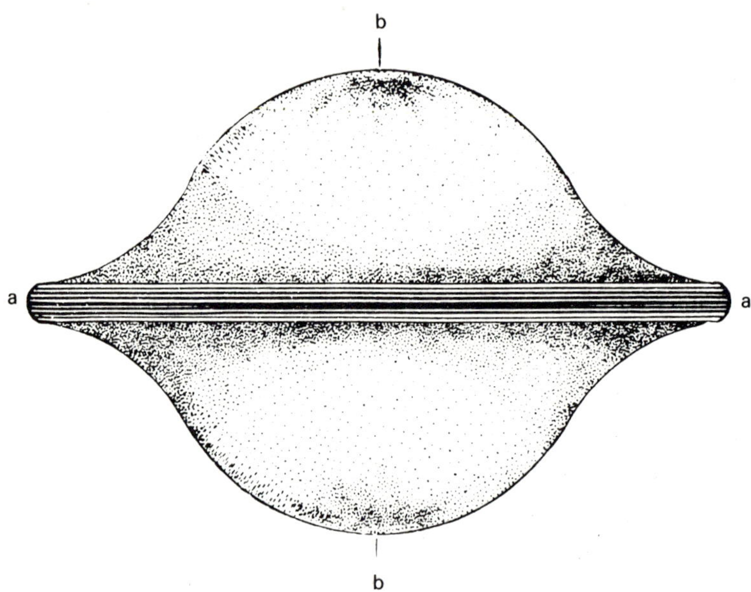

EARLY STAGE OF RING FORMATION

During the igneous era the rotation of the molten earth, supplemented by the expelling fires, carried all vapors, whether acqueous or metallic, to the terrestrial skies. These vapors took the direction along which the combined forces (heat and centrifugal energy) were strongest, i.e., to the equatorial heavens. In that day the young earth, could it have been viewed from the moon, would have presented approximately the form shown in figure 1.

The incipient rings are seen forming over the equator (a, a) and all vapors lifted into the polar skies (b, b) would, by their forced rotation, be carried toward the equator, as water is thrown from a turning wheel. (Rings must form at the equatorial plane.)

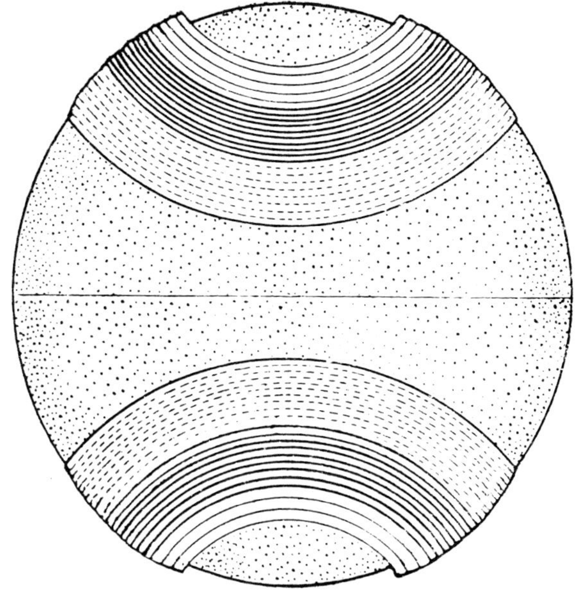

THE CLOSING SCENE
(Earth with Belts Capping the Poles)

This figure represents the earth stripped of its annular appendage and with its last lingering canopy suspended over the regions of both poles as vast clouds. Over the tropics and much of the temperature zones the vapors had become so thin that the clear sky could be seen at times and in places.

THE GLACIAL EPOCHS AND EDEN RUINS
Annular Snows the Only Competent Cause

The earth, now was a scene of death and almost boundless desolation. The unmistakable language of the geologic record is that there had just closed a long era of perpetual spring. The mammoth, mastodon, and multitude of other huge quadrupeds, whose giant remains are found in the world's stupendous wreck, fed upon the products of a tropical and semi-tropical earth. It was unmistakably a greenhouse world. The primitive elephant, and many of his congeners and contemporaries, fed in luxuriant forests and grassy plains, where now the glaciers and the arctic world are holding them in relentless grasp. A one time down rush of snow and ice built polar ice-caps. They are still there, there is no replacement process.

The author and daughter, Nancy, at the Petrified Forest in Arizona. Some claims are the Petrified Forest was buried for 200 million years under 3,000 feet of earth and rock. Does this seem possible?

The author and daughter Nancy (now Mrs. Steven Jore), just before take-off at the Palm Springs, California, Airport. Pilot Fred Hollister is already seated in the flight cabin. Mrs. Hollister (Peggy) is on board, and Mrs. Linderman (Mayvis) is taking the picture. She is an expert photographer. I enjoyed this flight and others as I occupied the copilot seat.

The author at the oil well Core Library located on the campus of the University of North Dakota at Grand Forks. This fine new department building bears the name of Dr. Wilson M. Laird, one-time department head.

Drilling deep in western Oklahoma's Anadarko Basin, this Tenneco wildcat was started in 1981 with a depth objective of 25,200 feet — a company record. It will require a full year to complete. Tenneco has large acreage holdings in this gas-prone area. *Photo by Tenneco Inc.*

The Williston Basin of the Dakotas and Montana represents a major investment for Tenneco. The company has spent nearly $220 million in this promising area for 1.5 million acres of exploratory land, drilling equipment and all the costly logistics of oil exploration. As a result, Tenneco has become a leading oil producer in the area. *Photo by Tenneco Inc.*

Tenneco's wildcat Arledge 1-17 discovered gas in Crockett County, West Texas, during 1981. Tenneco has enjoyed continued success in exploring the Permian Basin which underlies this area of Texas and southern New Mexico. *Photo by Tenneco Inc.*

This is a most outstanding oil well core taken from the #3 Oliver Bergstrom well, Haas Field, Bottineau County, North Dakota. This oil bearing core came from the interval of 3,968 to 4,011 feet deep. Completed July, 1980. The dark oil bearing section is called the Mission Canyon Formation. The break from impermeable Anhydritic Dolomite cap rock to Mission Canyon oil bearing limestone is instantaneous. This core sample supports my theory of annular development, not millions of years of slow erosion. The oil and oil bearing rock were placed first, then the impermeable cap rock on top sealing the oil until a drilling bit found it. Nothing could run down after placement. Mr. Rod Stoa, Curator of the Wilson M. Laird Core and Sample Library, North Dakota Geological Survey, Grand Forks, assisted in taking core photo and other NDGS pictures.

FIG. 15

The above is a standard rock bit. Drilling pipe is threaded onto the bit in thirty foot stands with water and drilling mud circulated down by pumping which brings cuttings to surface on the annulus side. If a bit must be changed at any depth, even 10,000 feet, all pipe must be removed, usually at every third joint for 90 foot stands. Two core samples, shown here, were taken from the 8,000 foot depth. A coring bit has no moving parts with commercial diamonds mounted which grind up the rock which is floated to surface the same as a rock bit. About 59 feet is all the average core barrel will hold, then it must be removed from the well, removing core rock. If more coring is required the core bit and barrel must be replaced in the well.

The above well was drilled by a major oil corporation from Tulsa, Oklahoma. This is a twin drilling site in the Fryburg Field, Billings County, North Dakota. Two wells use the same drilling pad near Theodore Roosevelt National Park, and are slant drilled under the Park. The wells are located near the Park Lookout on Highway 94 near Medora, North Dakota. By directionally drilling the wells are located out of the Park. Well H 818 shown here has a depth of 9,757 feet. The Lookout on Highway 94 gives one a beautiful view of the Badlands.

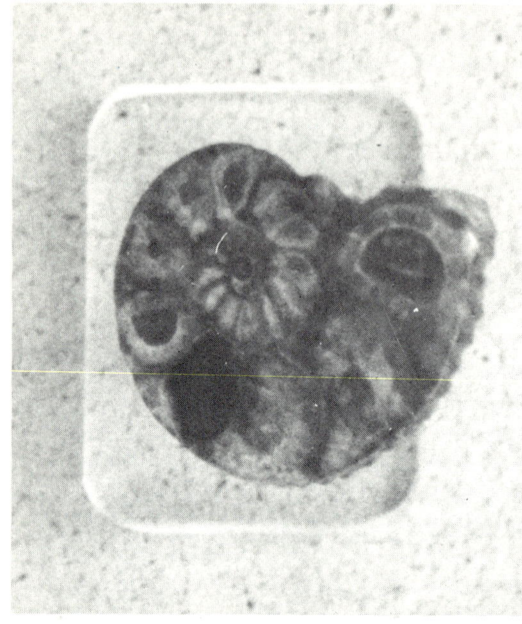

This is an Ammonite fossil shell now extinct. It is a coiled fossil shell (snail), assigned to the Mesozoic era. Many Ammonitic fossils were found in Big Horn Mountains of Wyoming and Montana, where this one came from. One was five inches in diameter.

In spending several days at the Jackson Lake Lodge, in the Grand Teton National Park, we took a rubber raft ride down the Snake River. Some of the most beautiful and interesting moraine formations I have seen are in the Snake River Valley along this river. In one turn of the river there was a side view of one about 50 feet above the river in the side bank. I do not attribute this to glaciation. This matrix was formed in a rather short time span in a fast moving high volume of water. In some places there was only a foot or two of soil on top of the moraine formation. Slow glaciation could not build this formation. Many small rocks are still rolling along on the river bottom. My question to the geologist is: How do we tell the difference if a rock is glaciated or just washed by water? In many cases who would accurately know?

Is there some great flaw in geological identity of many rock formations around the earth. Are they really sure all identified Precambrian, Cambrian, and upper Cambrian found around the globe are such? Some Precambrian is 16,000 feet deep. Is there yet another key as yet not found? A news item just in shows where a young man out for a walk on the family farm near Seaman, Ohio, found a trilobite fossil 10½ inches in size. The eighth-grader, Kevin Birchfield, found the fossil in a creek on the family farm. Most trilobite fossils are ¼ to 4 inches. Scientists have estimated the age of this fossil from 220 to 450 million years old. In old-school geology why would such a fossil be found on a farm in Ohio right on the surface? Trilobite fossils are usually associated with the Cambrian Age. Why should this fossil have been drifting about for hundreds of millions of years? Does this seem possible? In annular geology this fossil find could be accounted for by washing it to where it was by a great flood of water.

Are we to identify Cambrian formation mainly because it contains a specie of trilobite fossils? There are said to be 200 genera, containing 2,000 trilobite species. In various places on earth Cambrian rock is said to be from 7,000 to 40,000 feet thick. In its composition we may find sandstone, limestone, shales, slates, marble, quartzites, manganese, limonite, gold-bearing ores (Black Hills). Small wonder it is also conglomerate. If shallow or surface Cambrian has one of the trilobite fossils in its matrix, can we identify it solely on this basis? Because the earth has been tremendously overwashed by a great flood of water it very easily could be, various geological ages and fossils, at or near the surface are thrown together not related. They both could have become conglomerate in this process. Who would know which is which?

If we see a building with doors and windows are we to say it is a house? If we see some doors and windows in use, are we to assume it must be a house? The presence of a form of trilobite fossil in some shale or sandstone may not prove much.

In the loop of about 300 miles from Billings, Yellowstone, the Grand Tetons, Wapiti Valley, Cody, and the Big Horn Mountains, dozens of localized factors are attributed to the development of these worldwide features by geologists. It, thus, becomes enigmatic to assign local processes to a worldwide world building effort. I am more convinced than ever these grand convolutions had a single energy source, the collapse of the last water ring around the earth, with upheavals, hydro-forces beyond all comprehension tearing in at the earth's surface leaving it as we see it. Ocean augmentation thrust the continents higher and higher. Now, only, in our geological times, are the slow tedious forces of erosion beginning to do their work. Again, I must ask, could such massive worldwide geological structures in close proximity of each other have been developed independently from each other?

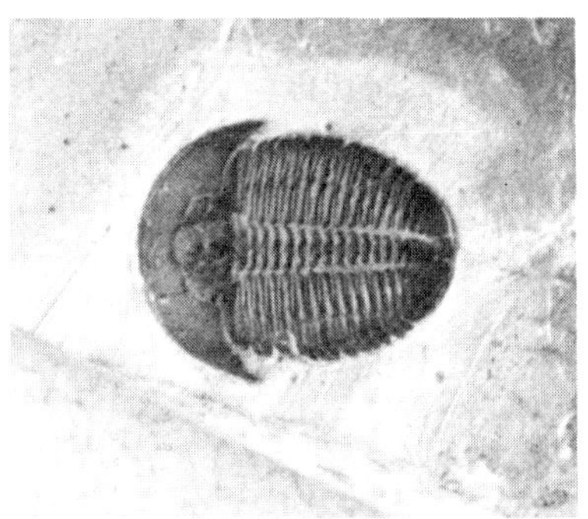

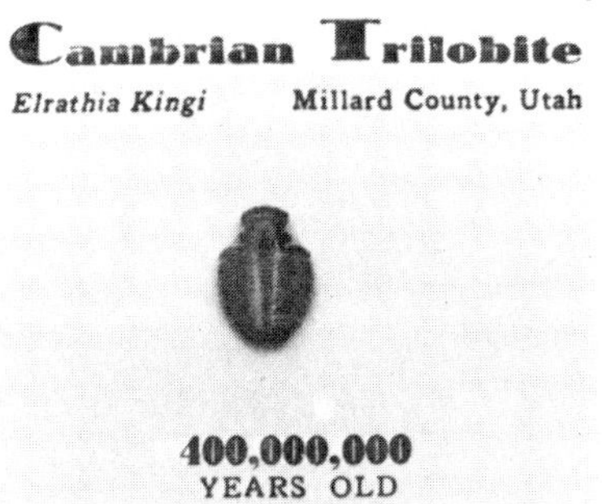

<table>
<tr><td>

The above is a Cambrian Trilobite I found at the Teton Gemstone shop in Jackson, Wyoming. The trilobite, now extinct, is 1 5/8 inches long, in the original matrix. The small specks around it are baby trilobite. The fossil is from Utah and appears to be a shale or limestone, thought to be of the Cambrian age.

</td><td>

Above is also a Cambrian Trilobite fossil from Utah. These are found near the surface. The assigned age of the fossil is not mine. I will not say how old they are at this time. They appear to have been washed over by a great flood of water in a short time span embedding them in the sedimentary matrix.

</td></tr>
</table>

In the study of geological development the matrix of the Cambrian trilobite, some Wyoming oil shale, and a mission canyon oil core sample from 8,000 feet deep in the Williston Basin, all appear about the same. They appear to have been washed over by an ocean of water. I have a beautiful pottery vase from the giftshop at the Northern Hotel in Billings, Montana. It has three natural colors, ivory, medium brown and dark red. It is made of Precambrian Grayson Shale from western Montana. The range of Cambrian, Precambrian sedimentation, if our assessment is correct, amazes us. Can this always be correct?

On a trip with my family, including daughter Nancy and husband Steven Jore, I made some very interesting studies of geology. The extinct trilobite with its babies nearby seem to be about like the extinct Mammoth of Siberia, with their young next to them and food still undigested in their stomach. Just east of Yellowstone I saw a Mammoth tusk on display thought to be similar to the Siberian Mammoth, which once roamed the central U.S.A. and Wyoming. Perhaps the downfall of one or more of the rings once around the earth caused the extinction of the Trilobite and also the Mammoth.

It is not my intention to cast a reflection on intelligent and respected geologists. Yet, I will ask some questions about much of the accepted geological theory of the last one hundred years. Perhaps much of it has become a contagion rather than a reflection of fact. To set the stage, let's start at Pompey's Pillar 23 miles east of Billings, Montana. Next go to the Red Lodge, going north into the higher elevations of the Bear Tooth Mountains, where we will see some large boulders six feet and more in diameter with a tree growing out of the top of some of them. The claim is these are Cambrian boulders 50 miles from a Cambrian source. As usual, they say a local glacier moved them to this area. They are found helter-skelter over a wide area with trees growing around them. A large volume of water could develop this arrangement, a glacier could not.

To forever have glaciation, with great heat and hydrothermal development, **and** prolific plant life all at the same time seems a major contradiction. Snow pack and ice heaps in mountain peaks and valleys is not true glaciation. These are too puny. The mighty gorge of the Yellowstone River with its upper and lower falls is called the Grand Canyon of the Yellowstone. Looking up from the base of lower falls, or standing on the rim of this canyon, we see a much higher surface regional elevation than any part of the river. This gorge was not cut by a tiny trickle of water through a crack in the earth millions of years ago. It was caused by the same omnipotent force that made the Grand Canyon, a great fall of water from an earthly ring system, in a short time span. Glaciation had little to do with its development. In much of this area it is claimed that glaciers slid into valleys damming up water to create the many features. Geologists also claim this happened repeatedly. One national publication postulates that such an ice sheet dammed a river on the Idaho-Montana border, creating a lake 2,000 feet deep, repeating this several times flooding into the Columbia River. I see no such action possible. This would require one miracle upon another.

There is a so-called Precambrian boulder weighing 500 tons estimated to have been pushed about 15 miles by a glacier, near Inspiration Point. Could a glacier do this? Perhaps, if the rock could get on top of the ice and later fall through. This brings to mind when many years ago I attended a picnic with my uncle and friends in one of the mountain arroyos out of Los Angeles, California. My uncle said to me; see those huge boulders (some of them many times as large as a car). He said, last time we were here those boulders were not here. There was a big flood in the area and they were washed in from miles away. This proves my point.

The term ice field build-up is used by many geologists for this area. There are giant petrified tree trunks in Yellowstone. When did these grow with a constant ice age?

Here are two formations in Zion National Park, Utah. If this picture were in color we would see one rock formation very dark reddish in color, and the other very light in color. The swirl formation indicates they were deveoped by an inveterate igneous agency, not erosion. The proximity factor tells us they are not built by slow tedious erosion. They are not metamorphosed either. A very great volume of water washed them clean of other debris and are now just as nature left them.